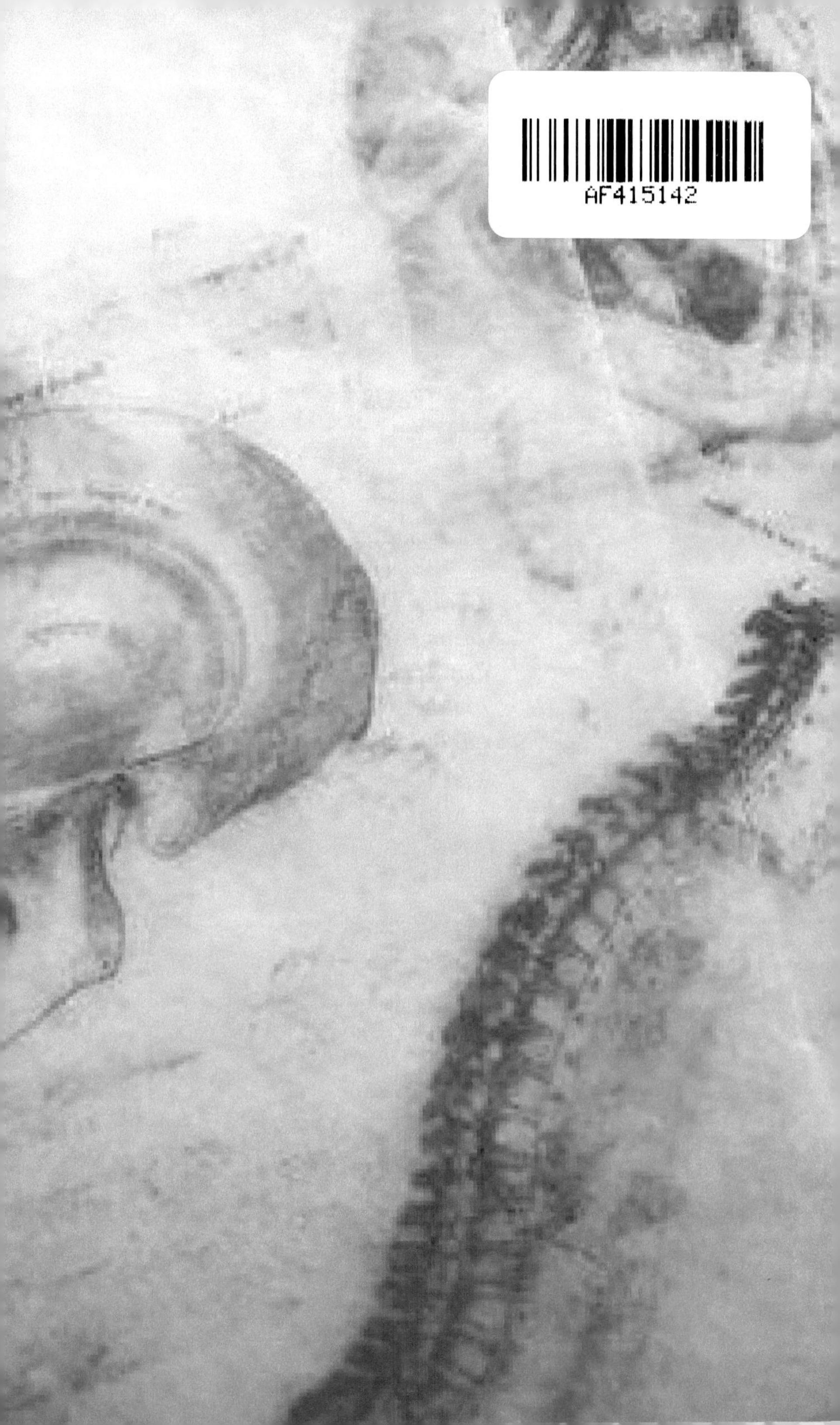
AF415142

U-N-I-VERSE
- VOLUME I -
JESUS

Featuring
Saleem
George Hopkins
Norman McDonald
Shavar Thompson
Imhotep Shabazz
Saleem bin Shasmuldeen

MITANNI PUBLISHING LLC

Copyright © 2021 Mitanni Publishing LLC

U - N - I - VERSE
- VOLUME I -
JESUS

TABLE OF CONTENTS

DEDICATION
The Seekers

FOREWARD

When I look at the Universe (Sun, Moon and Stars) I see something that is missing in humanity. That element – which is null and void – would provide a blueprint for us to follow that would propel the human families in the immediate direction of success.

That which I recognize in the Universe can be found the word itself. Let me establish, before I proceed, that the Physical Universe that we all see, live in, and submit to consist of the Sun, Moon, Stars and all that exist in between.

The Term "Universe" is made up of a prefix (Uni) and a single word (Verse). Prefix means before and *"Uni"* means *"One"*. While *"Verse"* is another term for *"Word"*. Now when you look at the word *"Uni-Verse"* as a sum of its parts you have *"One-Word"* and that "One Word" is to *"Be"*. *"Be"* means *"to exist (in a form of I Study Life Around Me (Islam))."*

The Universe was set in place to remind us to become one and/or exist as "One" under the banner of *Islam.*

Upon further investigation I was able to dissect the word even further, by breaking it down into sections and manifesting a paralleling between us and that which seems to be *"Out-of-this-world"* in a manner of speaking.

So, when I assayed the word *Universe,* I came up with U-N-I-Verse – meaning; " *You and I Verse or Versus"* and it added much more substance to the above said meaning. Where You and I (U-N-I) must become one; we have to unite in order to exist and verse the devil's uncivilized and wicked machinations, both mentally and physically

- *George Hopkins*

U-N-I-VERSE I: JESUS

PREFACE

Mitanni Publishing is proud to present the first installment in the "U-N-I-Verse" series. "U-N-I-Verse" is a platform for all "seekers" to share wisdom with the world. It is a series where writers from all religions, schools of thought and walks of life come together to discuss differences and similarities in order to shed light on and build upon common points of interest that may lead to Universal balance and harmony. This series is intended to unify the human family and realign all with the cosmic symphony that man's ego and pride has disrupted.

In Volume One of the series, we will be exploring the often-enigmatic character known as "Jesus", "Messiah", "Christ", "Yeshua", "Isa Ibn Maryam" and countless other names and epithets. One of the most monumental figures in human history, Jesus' life and teachings has been shrouded in mystery and confusion while still supplying ample light in regards to the distorted representations of his true teachings or Gospel (Injeel).

Amidst the confusion of his life and death and alleged crucifixion, the true teachings of Jesus have succumbed to the degeneration of time and yet and still, the little that remains, has been enough to transform the lives of the poor and destitute, the forlorn monk, the debased drug addict and countless other lives. The remnants of his teachings have inspired the innocent youth and the cunning man; the simpleton and the sage alike. This fact alone should attest to the greatness of his Gospel (Injeel) for if fractions of his teachings have led to the spiritual transformation of many human souls, surely that which he brought to his disciples, opposers, and strangers through Divine Inspiration must have been even greater. Nonetheless, people from all walks of life share an affinity for the man and the myth, born to the Virgin Mary known as: *"Jesus of Nazareth"*

- Saleem

INTRODUCTION

DEFINITION
universe (n.) *1580s,*
"the whole world, cosmos, the totality of existing things," from Old French univers (12c.), from Latin universum *"all things, everybody, all people, the whole world,"* noun use of neuter of adjective universus *"all together, all in one, whole, entire, relating to all,"* literally "turned into one," from unus *"one"* (from PIE root *oi-no- "one, unique") + versus, past participle of vertere "to turn, turn back, be turned; convert, transform, translate; be changed" (from PIE root *wer- (2) "to turn, bend").

SUPREME WISDOM
"21st letter in the English Alphabet "U"

"**You**" or "**Universe**" - When you wisdom (speak) your knowledge you bring forth understanding which is the completion of your Universe. The universe is everything-sun, moon and stars. They are celestial bodies. When **U–N–I–VERSE** we bring about a more glorious and higher elevation of understanding. You must be a part to make something whole and strong so it can't be destroyed by anything weak. You be me; I be you and together we are one Allah. U-N-I-verse all negativity.

You are the universe because you are the producer of the sun, moon and stars. The man, woman and child."

**- Supreme Wisdom
NGE (Nation of Gods and Earths)**

"Now around this time lived Jesus, a wise man. For he was a worker of amazing deeds and was a teacher of people who gladly accept the truth. He won over both many Jews and many Greeks. Pilate, when he heard him accused by the leading men among us, condemned him to the cross, (but) those who had first loved him did not cease (doing so). To this day the tribe of Christians named after him has not disappeared"

-Josephus (37-101AD

NAZARETH

JESUS OF NAZARETH
By Norman McDonald

Without regard to the beliefs or claims of any individual or group, it cannot be denied that Jesus of Nazareth is one of, if not the most influential people in human history. Even if one chooses to delineate the current numbered year date without reference to "Before Christ," or "Anno Domini," and instead use BCE for "Before Common Era," the epoch that creates that distinctive starting point is unequivocally the birth of Jesus of Nazareth. Many other faith-based calendars are clearly in use in their respective communions, e.g., the Hebrew and Muslims Calendars, however all scientific, medical, political and historical context is still with reference to the birth of Christ as its demarcation.

Calendars, and the indices of time are a very minor issue as compared to the identity of Jesus of Nazareth in the thinking and sincerely held beliefs of mankind. Sometimes to the degree of a life-or-death determination, or one's ability to live peacefully in a certain global region, is one's profession of religious belief, and in particular, the identification of Jesus of Nazareth. Misinterpretation of the tenets of any given faith walk, and in particular, who Jesus of Nazareth is or was and what he represents to any given group has as its ultimate outcome in the historical strife and in most cases armed conflict of parties of different beliefs. The question remains, is this in line with any of his teachings?

The stated objective of this volume of "U-N-I-VERSE: Vol. I: "Jesus," is an exposition of unity, versus the abject separation and hostility that has historically been treated with this subject. We need to revisit none of the specific historical examples of violent and lethal consequences of the difference of opinions on the identity of Jesus of Nazareth. There is no shortage of wrongdoing "in the name," but we are left with the question of why a set of beliefs about the person, or the divinity of Jesus, has and is the cause of so much human suffering. Especially when his message was, as it has been recorded of peace and love.

In the Gospel accounts of the teachings of Jesus of Nazareth, he speaks of loving one's neighbor more than one's self, and even to "bless those who curse you." It remains a mystery how anyone of good conscience could interpret those ideas as "kill or torture anyone who doesn't subscribe to the exact same conviction, or in the exact same building." It makes even less sense that a group would conceive of setting fire to the most visibly recognizable symbol of Jesus of Nazareth in such a location so as to intimidate another group of people, presumably others of the same faith, in the interest of some flavor of "Supremacy." Clearly, in the teachings of Jesus, he was clear that there were to be no divisions of supremacy among people. In his time, it was certainly a normal function of Roman and Jewish societies that there be a clearly defined social order, or "caste." This attitude was never more evident than in the distinction between "chosen" and not chosen.

Whether or not one considers oneself a "Christian," the teachings of Jesus of Nazareth, even if not taken as being of divine origin, make sense as they relate to how we accord one another, and how we engender interpersonal relationships. A social order of human beings being necessary to our survival, the concept of forgiveness versus revenge, grudge holding or vengefulness is one that has physical as well as emotional benefits. The cumulative stress of negative and acrimonious attitudes creates in the human body the release of corticoid stress markers in the bloodstream, that have injurious effects on cardiac and cerebral wellbeing. Simply stated, the dwelling upon of hatred, revenge or angst has the cumulative effect of weakening your heart and mind to the point of fatal illnesses. The opposite is true of the old cliché, "only the good die young."

As any discussion or examination of the universal significance of the person or divinity of Jesus of Nazareth with respect to the cosmos, one would certainly have to direct any analysis into spiritual rather than temporal or secular aspects of his life and death. Various eastern faiths, most especially the Shinto faith of Japan, invests in any animate or inanimate being a soul or spirit. Clearly the beliefs of the Native American Indian do likewise, as geological features and animal life are indwelled with spiritual entities. Mankind for thousands of years has contemplated what is beyond our own sphere of habitable territory, and what are our stations and influences in the Cosmo at large.

As society becomes more and more aware of our effect on our planets, a logical discussion follows about what are the implications of what we represent in the grander scheme of the physical Universe, and beyond. That concept of beyond the physical Universe is clearly the one of most challenging of queries that mankind has ever struggled with.

As members of what we strive to consider ourselves as an enlightened world society, the divisions that have been created with respect to the identity of Jesus of Nazareth, as well as numerous articles of faith are counterintuitive and counterproductive to that end. It is not enough to *"walk a mile in a man's shoes,"* to empathize with his beliefs, much less respect them. It takes real effort, and an open mind to go beyond our own historical and personal prejudices and grudges to appreciate another's sense of belief and place in the world or in the Cosmo.

There are many very positive and benevolent aspects of the *"democratic experiment"* that America has engendered, and the attitude of celebration of diversity of culture and diversity of opinion is at the very top of that list.

To quote former U.S. President Barak Obama, *"One has the right to one's own opinions, but has no right to his own facts."*

There is no shortage of Hollywood movies that expound on the theme of how mankind is doomed to mutually assured destruction if we cannot resolve our tendencies for violent conflict over petty difference in opinions or beliefs.

To quote from a significant event in the earthly ministry of Jesus, as recorded in the Gospel account of Saint John, he reached out to a woman who was in the culture of his society ritually unclean, at a well in Samaria. Many believe that this was the beginning of his outreach to people of all faiths and stations in life, and not just to the chosen or elect. Without regard to what our spiritual beliefs are about the divinity or spiritual place of Jesus as either the Messiah or Christ, prophet or just a teacher, his example is not diminished by the interpretation of his identity.

In a truly enlightened society, we learn ways to accord each other's differing views with respect and dignity. We can still engender compassion for someone who believes differently from us and engage in healthy and respectful dialogue that serves to bring us closer rather than creating and magnifying those divisions. In my belief, that is the qualifying value of any analysis of who or what Jesus of Nazareth stands for.

WHAT IS PARTHENOGENESIS?

(/ˌpɑːrθɪnoʊˈdʒɛnɪsɪs, -θɪnə-/;[1][2] from the Greek παρθένος parthenos, "virgin", + γένεσις genesis, "creation"[3])

Parthenogenesis is a natural form of asexual reproduction in which growth and development of embryos occur without fertilization. In animals, parthenogenesis means development of an embryo from an unfertilized egg cell and is a component process of apomixes.

Gynogenesis and Pseudogamy are closely related phenomena in which a sperm cell or pollen initiates the development of the egg cell into an embryo but makes no genetic contribution to the embryo. The rest of the cytology and genetics of these phenomena are mostly identical to that of parthenogenesis.

The word parthenogenesis is sometimes used inaccurately to describe the reproduction process in hermaphroditic species that can reproduce by themselves because they contain reproductive organs of both sexes in a single individual's body. However, these species still use fertilization and hence the term parthenogenesis wouldn't be applicable.

Parthenogenesis is a process that occurs naturally in many plants, some invertebrate animal species (including nematodes, water fleas, some scorpions, aphids, some bees, some Phasmida and parasitic wasps) and a few vertebrates (such as some fish, amphibians, reptiles and very rarely birds). This type of reproduction has been induced artificially in a few species including fish and amphibians.

Normal egg cells form after meiosis and are haploid, with half as many chromosomes as their mother's body cells. Haploid individuals, however, are usually non-viable, and parthenogenetic offspring usually have the diploid chromosome number. Depending on the mechanism involved in restoring the diploid number of chromosomes, parthenogenetic offspring may have anywhere between all and half of the mother's alleles. The offspring having all of the mother's genetic material are called full clones and those having only half are called half clones. Full clones are usually formed without meiosis. If meiosis occurs, the offspring will get only a fraction of the mother's alleles.

Parthenogenetic offspring in species that use either the XY or the X0 sex-determination system have two X chromosomes and are female. In species that use the ZW sex-determination system, they have either two Z chromosomes (male) or two W chromosomes (mostly non-viable but rarely a female), or they could have one Z and one W chromosome (female).

GLOSSARY

Parthenogenesis reproduction from an ovum without fertilization, especially as a normal process in some invertebrates and lower plants.

Asexual (of reproduction) not involving the fusion of gametes:

Apomixes asexual reproduction in plants, in particular agamospermy. Often contrasted with amphimixis.

Gynogenesis a form of parthenogenesis, is a system of asexual reproduction that requires the presence of sperm without the actual contribution of its DNA for completion.

Pseudogamy refers to aspects of reproduction. It has different (but related) meanings in zoology and in botany.

Cytolgy the branch of biology concerned with the structure and function of plant and animal cells.

Meiosis a type of cell division that results in four daughter cells each with half the number of chromosomes of the parent cell, as in the production of gametes and plant spores: Compare with mitosis.

Haploid (of a cell or nucleus) having a single set of unpaired chromosomes. Compare with diploid.

REFERENCES

1. *"parthenogenesis". Merriam-Webster Dictionary.*

2. *"parthenogenesis - definition of parthenogenesis in English from the Oxford dictionary". OxfordDictionaries.com. Retrieved2016-01-20.*

3. *Liddell, Scott, Jones. γένεσις A.II, A Greek-English Lexicon, Oxford: Clarendon Press, 1940. q.v.*

4. *"Female Sharks Can Reproduce Alone, Researchers Find", Washington Post, Wednesday, May 23, 2007; Page A02*

5. *Halliday, Tim R.; Kraig Adler (eds.) (1986). Reptiles & Amphibians. Torstar Books. p. 101. ISBN 0-920269-81-8. Cite uses deprecated parameter /coauthors= (help)*

6. *Walker, Brian (2010-11-11). "Scientists discover unknown lizard species at lunch buffet". CNN. Retrieved 2010-11-11.*

7. *Savage, Thomas F. (September 12, 2005). "A Guide to the Recognition of Parthenogenesis in Incubated Turkey Eggs".Oregon State University. Retrieved 2006-10-11.*

8. *^ Jump up to:ª ᵇ Booth, W.; Johnson, D. H.; Moore, S.; Schal, C.; Vargo, E. L. (2010). "Evidence for viable, non-clonal but fatherless Boa constrictors". Biology Letters. 7 (2): 253–256.doi:10.1098/rsbl.2010.0793. PMC 3061174 .PMID 21047849.*

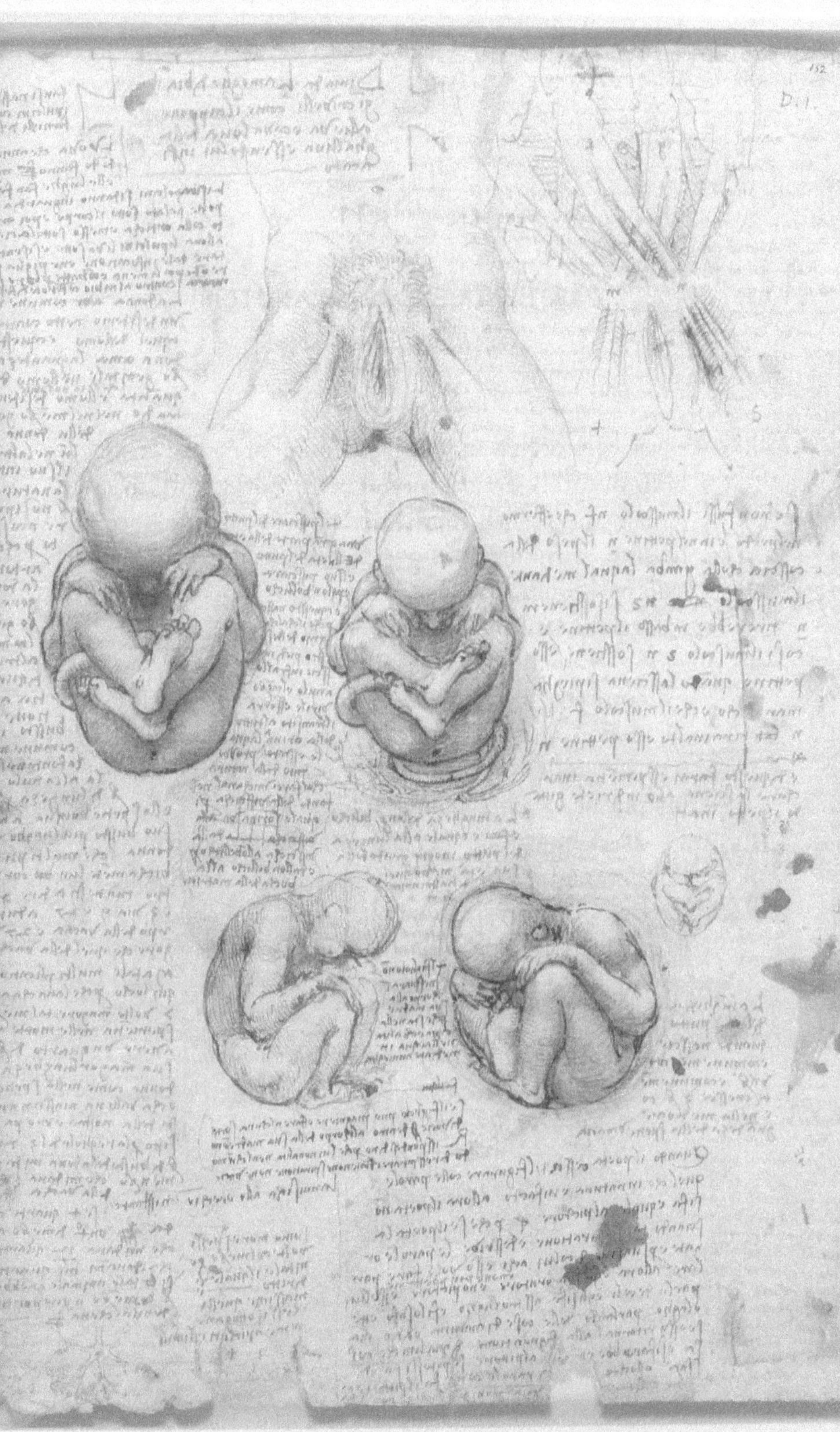

PARTHENOGENESIS (BE AND IT IS)

by Imhotep Shabazz

Jesus' first and most profound miracle was his birth of to his virginal mother. It is now known that female wasps, fish, birds, and lizards can produce healthy offspring without sexual intercourse or external fertilization, but what about humans? Are natural human virgin births possible?

Scientist say "Yes" - in theory. A number of extraordinary events would have to occur in close succession, and the chances of these all happening in real life are practically zero however. Not impossible but practically impossible.

A "practicality" is in no way an "impossibility" however and surely the one who creates and maintains the laws of gravity, physics, magnetism and child birth is capable of shifting them to suit his majesty. Allah says in the Quran of Jesus' birth:

"The similitude of Jesus before Allah is as that of Adam; He created him from dust, then said to him: "Be". And he was."
-Surah 3:59

To shift the laws of nature to make way for a new creation or the adaptation and evolution of a creature is in no way difficult for that which brought an entire Universe into existence from nothing. However, for those who have monitored nothing but the fixed laws of nature, it is hard to perceive the bending and adjusting of the laws of physics without imagining catastrophe because of the fine tuning of the universe. Back to Science...

Scientist say, for a virgin to get pregnant, one of her eggs would have to produce, on its own, the biochemical changes indicative of fertilization, and then divide abnormally to compensate for the lack of sperm DNA. These two events occur in the eggs or egg precursor cells of one out of every few thousand women. The egg would also need to be carrying at least two specific genetic deletions to produce a viable offspring.

An egg will only start dividing once it senses a spike in cellular calcium. This typically takes place as a result of a sperm's entry to the egg during fertilization. But if the egg happens to experience a spontaneous calcium spike, it will start reacting as if it's been fertilized. A defective sperm that lacks DNA can produce a spurious calcium spike. In the lab, scientists can coax unfertilized eggs into beginning the post-fertilization process by simply injecting them with calcium.

"Are you a more difficult creation or is the heaven?
Allah constructed it."
-Surah 79:27

Once fertilization occurs, an egg can complete the final stage of a cell division known as meiosis II, during which it loses half of its genetic material to make room for the sperm's DNA. If there is no sperm however, as is considered to be the case with Jesus, then each half of the divided egg cell will end up short, and both will die. In order for our "Immaculate Conception" to advance, the fertilized egg must, therefore, not complete meiosis.

Both the calcium spike and the division mistake could occur as the result of random dysfunctions or genetic defects. Assuming they do, the egg cell may then begin the process of "parthenogenesis," or virginal development. When this happens to an egg-precursor cell, it can give rise to a tumor made up of many different types of tissue—liver, teeth, eye, and hair, for example.

"That is Jesus, the son of Mary - the word of truth about which they are in dispute. It is not [befitting] for Allah to take a son; exalted is He! When He decrees an affair, He only says to it, "Be," and it is."
-Surah 19:34-35

Parthenogenesis in humans never produces sustainable embryos, however, because unfertilized eggs lack specific instructions about gene expression from the sperm. In general, our cells have two functional copies of each gene—one inherited from the mother and one from the father. For some genes, however, only one copy is ever used, while the other remains dormant. Some of the signals for which copies should be turned off come from the sperm cell. So, if there's no sperm, certain genes will be overexpressed, and the "embryo" will die when it is only about five days old.

There's a way around this problem, too. By eliminating a pair of maternal genes, a Japanese team was able to create, via parthenogenesis, a viable baby mouse that was seemingly unaffected by its lack of paternal imprinting. Although the scientists engineered these changes in the lab, there's at least a theoretical possibility that this could happen spontaneously via random gene deletions.

So, while it's possible for a human baby to be born of a virgin mother, it's very, very unlikely: These two genetic deletions might each have a one in 1 billion chance of occurring, and that's not counting the calcium spike and division problem required to initiate parthenogenesis in the first place.

According to a 1995 report in the journal Nature Genetics, a mother brought her infant boy to the doctor after noticing that his head was developing abnormally. When doctors analyzed his blood, they found something truly bizarre: Despite his anatomically male features, the boy's blood cells were entirely female, consisting only of genetic material from his mother. Some of his other cells—such as those found in his urine—were normal, consisting of a combination of both maternal and paternal DNA. No one knows exactly how this occurred, but the best guess is that immediately after being fertilized, one of his mother's eggs fused with a neighboring unfertilized egg that was dividing parthogenetically. This gave rise to a boy who was considered half-parthenogenetic, since approximately half of his cells were derived from a "faux" conception, containing no remnants of his father's DNA.

On August 2, 2007, after much independent investigation, it was revealed that discredited South Korean scientist Hwang Woo-Suk unknowingly produced the first human embryos resulting from parthenogenesis. Initially, Hwang claimed he and his team had extracted stem cells from cloned human embryos, a result later found to be fabricated. Further examination of the chromosomes of these cells shows indicators of parthenogenesis in those extracted stem cells, similar to those found in the mice created by Tokyo scientists in 2004. Although Hwang deceived the world about being the first to create artificially cloned human embryos, he did contribute a major breakthrough to stem cell research by creating human embryos using parthenogenesis. The truth was discovered in 2007, long after the embryos were created by him and his team in February 2004. This made Hwang the first, unknowingly, to successfully perform the process of parthenogenesis to create a human embryo and, ultimately, a human parthenogenetic stem cell line.

It appears that pragmatic minds hell-bent on disproving any sort of Universal programming insist on finding ways the miraculous cannot occur. However, theories never rule out the possibility of exceptions to rules, as is the case with many birth defects and other Universal Phenomena that cannot be explained by the human mind.

Yet and still, there are those who are wise enough to admit when they've arrived at a point where human comprehension no longer functions, and they are forced to admit that there are possibilities. For some, Science is needed to explain "miracles" (or discredit them) for others intuition is enough to remind a man that he's not sure how he himself came into existence let alone how things before his existence came to be as finely tuned and cooperative as the Universe and all contained within. (Evolution explains a phenomenon of things already in existence. adaptation is a fact. However, it is not as simple as saying "evolution" brought the Universe into existence. Most men die before their theories of how life sprang from "nothing" are discredited.)

To add to this point of theory, hypothesis and conjecture; one of the most World-renowned Scientist, philosophers and thinkers of our time proposed a theory referred to as "The Static Universe" or "The Stationary universe."

A static universe, also called a "stationary" or "Einstein" universe, was a model proposed by Albert Einstein in 1917. It was problematic from the beginning. Edwin Hubble's discovery of the relationship between red shift obliterated it by completely demonstrating that the universe is constantly expanding. Meaning, the Universe was not infinite and eternal as first believed by those who proposed the same theory of energy having no beginning or end and not being able to be created or destroyed.

And that is one of tens of thousands of Scientific Theories proposed to the world as fact based upon *empirical* evidence only to be debunked later on in life. The one reality that can never be debunked is your existence. Hence, searching the world and the Universe for answers is less fruitful than searching the self. The heart. The source of comprehension and consciousness. You can spend hours reading books about things that may or may not be true or you could spend a few minutes in meditation contemplating your existence and doing good deeds. This is the way of Jesus. Many times, he explained to his teachers that they could not teach him from books for they didn't know themselves nor their creator.

THE INFANCY GOSPEL OF THOMAS

6:1 There was an instructor named Zacchaeus standing off to the side who heard Jesus say these things to his father. And he was amazed that he was speaking such things, though just a child. 2 After a few days he approached Joseph and said to him,

"You have a bright child with a good mind. Come, let me have him that he may learn to read, and through reading I will teach him everything, including how to greet all the elders and to honor them as his ancestors and fathers, and to love children his own

age." 3 And he told him all the letters from Alpha to Omega, clearly and with great precision.

But Jesus looked the Instructor Zacchaeus in the face and said to him, "Since you do not know the true nature of the Alpha, how can you teach anyone the Beta? You hypocrite! If you know it, first teach the Alpha, and then we will believe you about the Beta." Then he began to question the teacher sharply about the first letter, and he was not able to give him the answers.

4 And while many others were listening, the child said to Zacchaeus, "listen, teacher, to the arrangement of the first letter of the alphabet; observe here how it has set patterns, and middle strokes which you see collectively crossing, then coming together, and proceeding upward again till they reach the top, so that it is divided into three equal parts, each of them fundamental and foundational, of equal length. Now you have the set patterns of the Alpha."

It is as if the young Jesus is saying, "If you do not know the beginning, the fundamental principles, the Alpha that leads to the Beta, the source that initiates creation; if you do not even know the structure of the Universe, its formation and how it was formed as well as its metaphysical secrets, how are you going to teach me about life or myself with adulterated books written by mere humans with flawed logic, ulterior motives and vain ambitions?"

On the other hand, maybe the scientific facts you speak of are true. But will knowing them make me a better person, henceforth creating a better world, and a higher chance for spiritual success? There is much arrogance in knowledge and the most shameful aspect of this is, it is regurgitated data and not internal enlightenment. If knowledge worsens the condition of the human spirit, it is counter-conducive and utterly useless.

Maybe Parthenogenesis can one day explain an *"Immaculate Conception"*, maybe not. The creator reveals through inspiration that indeed a miracle and a sign was born for people to witness but they belied that sign just as they belied the signs of the rest of the prophets. Maybe we seek to uncover external mysteries that have no bearing on our internal state. Maybe the man was never the focus, maybe it was his message.

Peace and Blessings

THE MAN, THE MYTH, THE LEGEND
By Shavar L. Thompson

To some people Jesus represents God in human form. To others Isa is a mortal messenger chosen to further God's message of conducting oneself in a holy manner while strictly adhering to God's commandments (law). Both arguments are outdriven by Jesus' conduct while on Earth alone. Too much focus is exhausted trying to distinguish factors that cause folks to circle around the standards of comportment that holy men must live by. To me Jesus represents how we should strive to conduct ourselves daily, and how motivated we should be to perfect our knowledge of the books we attest as truth.

In this profit -driven era where morals, values, and character are sacrificed for personal gain; a lot of us (including myself) become blinded by the opportunity to obtain possessions. Resulting in selfishness, murder, conceit, deception, hate, pollution and division. Personal gain interferes with one's connection to the spirit, unconsciously creating a self-satisfaction seeking and impulsive race; Putting us in direct conflict with Jesus and his teachings.

Jesus spent his entire adult life on earth as a selfless provider, healer, prophet and servant.

Spreading the message of the importance of recognizing oneness with all beings. Encouraging us to cherish the environment we all share. Jesus displayed unconditional stewardship and charity to the masses under extreme amounts of duress. He encouraged us to always help those in need, and to hold a special place of compassion for the sick, poor, and elderly. Jesus told us of a duty to remember aiding others over one's own desires. To always strive to be generous., polite, and to act with integrity.

Jesus has transformed in meaning in this age; from a man chosen to suffer for my sins, to a prophet who has the demeanor that must be modeled if one wishes to attain a place in paradise. In addition, one can attain true bliss while in the devil's playground, and absolute harmony with all living things on earth before perishing. One of the most significant facts of Jesus' travels is that he lived a wholesome life his entire time on Earth. He never deviated from positivity, aiding and nurturing other beings. Humbling himself enough to clean the feet of the elderly, poor, and sick. Finding refuge amongst the outcasts of society and believing in the potential of all humans even in their worst state.

Jesus didn't accept the norm of valuing property and self-prosperity over activism and philanthropic ideals. He never put his own agenda ahead of investing time in perfecting the mind states of people. Jesus denied materialism and imperialism when he encouraged people to realize there are no differences among us. To recognize that as humans we need to appreciate the glory of having one another in this perfect moment. Not creating caste systems where few thrive while the majority suffers.

When Jesus commanded us to aide, encourage, and show respect for the poor, he meant for us to realize our own individual insignificance in the big scheme of our reality. To grasp that many things the Creator loves are sacrificed every day to allow our continuation.

By understanding the above, we would be able to recognize that a lot of the time we waste being judgmental and creating outcasts is misplaced. Most people think literally poor, causing them to become oblivious to the fact that the poor exist in many forms. We live in times where folks are spiritually poor (No disrespect intended). Jesus signifies to me a duty to unconditionally help those who suffer from any of the above conditions. Our states of mind should be hauling our neighbors over the hump. We should have a commitment to overlook race, financial status, social standing and physical handicaps; and replace them with compassion. The desire to achieve the best results for everyone has to become our nation's biggest priority.

Jesus taught us not to value profit or self-fame over the welfare of our neighbors and friends. Many mighty nations before ours that became addicted to possessions of the material world have failed. Inevitably we will face the same fate if we don't stop committing the materialism and start perfecting the comportment that Jesus intended us to actualize through self-demonstration.

In conclusion, Jesus means that I must strive every day to be the most caring, loving, helpful, behaved, faithful, joyous, etc... servant to mankind I can possibly be. Jesus means that I must lose the desire to please peers and find significance in works that God finds inspiring. Jesus means that I will have to continuously struggle in this lifetime while doing God's work. Jesus means that I must not be deterred from positive actions in the face of persecution. Unbendable faith, unwavering hope, unlimited praise, and undeterred servitude come to mind when I try to define Jesus.

THE BABYLONIAN TALMUD

תַּלְמוּד

There are only a few clear references to Jesus in the Babylonian Talmud, a collection of Jewish rabbinical writings compiled between approximately A.D. 70-500. Given this time frame, it is naturally supposed that earlier references to Jesus are more likely to be historically reliable than later ones. In the case of the Talmud, the earliest period of compilation occurred between A.D. 70-200. The most significant reference to Jesus from this period states:

"On the eve of the Passover Yeshua was hanged. For forty days before the execution took place, a herald ... cried, "He is going forth to be stoned because he has practiced sorcery and enticed Israel to apostasy."

THE SOLAR STORY OF JESUS
By Kevin Lomax

To me, the story or stories of Jesus are symbolic or allegorical in nature. In this short essay I will be giving my views on Jesus and examples to support my views. Some of the stories you may hear about Jesus are based off of our solar system. Personified for us to better understand the zodiac signs.

It is known throughout history that Jesus is the Son but what son? He was changed from S.U.N. to S.O.N. to better fit man. In our study of the solar system Venus is referred to as the Morningstar. Its shines in the dawn as the brightest light in the sky. The term Lucifer is applied to Venus in regards to the way Venus shines. it is the first thing bearing light in the dawn. The term Lucifer is a Latin term for Light-Bearer. In orbit as the Morningstar heralds dawning of the Sun (Jesus), Venus (Lucifer) shines brightly, almost as a challenge to the Sun (Jesus). If we turn our attention to the book of Matthew chapter 3 verses 9-10, it shows a personified example of that challenge. I quote,

"And he said to him, all these things I will give you if you will fall down and worship me." Then Jesus said to him, *"away with you Satan!"*

When the Son Jesus brakes through the horizon it's the shine of Venus Lucifer, and Venus Lucifer is sent into darkness.

The New Testament, in my eyes, is a reflection of the Hebrew mythology. Our ancestor's study of the solar system has been distorted in a consequential way throughout the times. When they spoke of the "Sun" they meant just that and not a person (son). When you speak of the Crucifixion of Jesus on the cross that is referring to the crossing of the Sun in our constellation with Cygnus (the Swan) which is located with the Capricorn. The other two that were crucified alongside the personified Jesus refers to the equinoxes. See the one on the right is the Vernal equinox and the one on the left is the Autumnal equinox. they are called this because of daylight savings where it is as if they still time from each other.

The Resurrection of Jesus is based on December 22nd, the beginning of the winter where the Sun would seem as if it has died. Our ancestors had no knowledge of the 3-day solstice. Where the Sun was at its lowest point. So, they would perform rituals and sacrifices for the Sun to rise again. So, when the Sun would rise again not because of the rituals and sacrifices but because the Sun will start traveling North again and by December 25th the Sun would start to shine again. Some priest would say that December 25th is the day Jesus rose from the grave.

THE TALMUDIC VIEW OF JESUS

The Talmud teaches that Jesus Christ was illegitimate and was conceived during menstruation; that he had the soul of Esau; that he was a fool, a conjurer, a seducer; that he was crucified, buried in hell and set up as an idol ever since by his followers.

The following is narrated in the Tract Kallah,

1b: *"Once when the Elders were seated at the Gate, two young men passed by, one of whom had his head covered, the other with his head bare. Rabbi Eliezer remarked that the one in his bare head was illegitimate, a mamzer. Rabbi Jehoschua said that he was conceived during menstruation, ben niddah. Rabbi Akibah, however, said that he was both. Whereupon the others asked Rabbi Akibah why he dared to contradict his colleagues. He answered that he could prove what he said. He went therefore to the boy's mother whom he saw sitting in the market place selling vegetables and said to her: 'My daughter, if you will answer truthfully what I am going to ask you, I promise that you will be saved in the next life.' She demanded that he would swear to keep his promise, and Rabbi Akibah did so—but with his lips only, for in his heart he invalidated his oath. Then he said: 'Tell me, what kind of son is this of yours'? To which she replied: 'The day I was married I was having menstruation, and because of this my husband left me. But an evil spirit came and slept with me and from this intercourse my son was born to me.' Thus, it was proved that this young man was not only illegitimate but also conceived during the menstruation of his mother. And when his questioners heard this they declared: 'Great indeed was Rabbi Akibah when he corrected his Elders'! And they exclaimed: 'Blessed be the Lord God of Israel who revealed his secret to Rabbi Akibah the son of Joseph' "!*

"Rabbi Eliezer said to the Elders: 'Did not the son of Stada practice Egyptian magic by cutting it into his flesh?' They replied: 'He was a fool, and we do not pay attention to what fools do. The son of Stada, Pandira's son, etc.' " as above in Sanhedrin, 67a.

This magic of the son of Stada is explained as follows in the book Beth Jacob, fol. 127 a:

"The Magi, before they left Egypt, took special care not to put their magic in writing lest other peoples might come to learn it. But he devised a new way by which he inscribed it on his skin, or made cuts in his skin and inserted it there and which, when the wounds healed up, did not show what they meant."

"If all the things he did had prospered, if he had rebuilt the Sanctuary in its place, and had gathered together the dispersed tribes of Israel, then he would certainly be the Messiah...But if so far he has not done so and if he was killed, then it is clear he was not the Messiah whom the Law tells us to expect. He was similar to all the good and upright rulers of the House of David who died, and whom the Holy and Blessed Lord raised up for no other reason but to prove to many, as it is said (in Dan. XI, 35): And some of them who understand shall fall, to try and to purge them and to make them white, even till the end of time, because the appointed time is not yet.

Daniel also prophesized about Jesus the Nazarene who thought he was the Christ, and who was put to death by the judgment of the Senate: (Dan. V.14): ...and the robbers of thy people shall exalt themselves to establish the vision; but they shall fail. What could be plainer? For all the Prophets said that the Christ would set Israel free, would bring it salvation, restore its dispersed peoples and confirm their laws. But he was the cause of the destruction of Israel and caused the rest of them to be dispersed and humiliated, so that the Law was changed and the greater part of the world was seduced to worship another God.

Truly no one can understand the designs of the Creator, nor are his ways our ways. For all that has been built up by Jesus the Nazarene, and by the Turks who came after him, tend only to prepare the way for the coming of Christ the King, and to prepare the whole world equally for the service of the Lord, as it is said: For then I shall give a clean mouth to all peoples that all may call upon the name of the Lord, and bow down in unison before him. How is this being accomplished?

Already the whole world is filled with the praise of Christ, the Law and the Commandments, and his praises have spread to far distant lands and to peoples whose hearts and bodies are uncircumcised. These discuss with one another about the Law that was destroyed—some saying that the commandments were once true, but have ceased to exist; others that there is a great mystery about it, that the Messiah-King has come and that their doctrine has revealed it. But when the Christ truly comes and is successful, and is raised up and exalted, then everything will be changed and these things will be shown to be false and vain."

Hilkoth Melakhim (IX, 4)

ISA IBN MARYAM

Quran Tafsir by Ibn Kathir

After Allah, the Exalted, mentioned the story of Zakariya, and that He blessed him with a righteous, purified and blessed child even in his old age while his wife was barren, He then mentions the story of Maryam. Allah informs of His granting her a child named `Isa without a father being involved (in her pregnancy). Between these two stories there is an appropriate and similar relationship. Due to their closeness in meaning, Allah mentioned them here together, as well as in Surahs Al `Imran and Al-Anbiya'. Allah has mentioned these stories to show His Servants His ability, the might of His authority and that He has power over all things. Allah says,

"And mention in the Book, Maryam,"

She was Maryam bint `Imran from the family lineage of Dawud. She was from a good and wholesome family of the Children of Israel. Allah mentioned the story of her mother's pregnancy with her in Surah Al `Imran, and that she (Maryam's mother) dedicated her freely for the service of Allah. This meant that she dedicated the child (Maryam) to the service of the Masjid of the Sacred House (in

Jerusalem). Thus, they (Zakariya, Maryam's mother and Maryam) were similar in that aspect.

"So her Lord (Allah) accepted her with goodly acceptance. He made her grow in a good manner."
3:37

Thus, Maryam was raised among the Children of Israel in an honorable way. She was one of the female worshippers, well-known for her remarkable acts of worship, devotion and perseverance. She lived under the care of her brother-in-law, Zakariya, who was a Prophet of the Children of Israel at that time. He was a great man among them, whom they would refer to in their religious matters. Zakariya saw astonishing miracles occur from her that amazed him.

Every time Zakariya entered the Mihrab to (visit) her, he found her supplied with sustenance. He said:

"O Maryam! From where have you got this" She said, "This is from Allah." Verily, Allah provides to whom He wills, without limit." 3:37

It has been mentioned that he would find her with winter fruit during the summer and summer fruit during the winter. This has already been explained in Surah Al ` Imran. Then, when Allah wanted to grant her His servant and Messenger, `Isa, one of the five Great Messengers.

"She withdrew in seclusion from her family to place facing east." 19:16

This means that she withdrew from them and secluded herself from them. She went to the eastern side of the Sacred Masjid (in Jerusalem). It is reported from Ibn `Abbas that he said, "Verily, I am the most knowledgeable of Allah's creation of why the Christians took the east as the direction of devotional worship. They did because of Allah's statement,

"When she withdrew in seclusion from her family to a place facing east.) Therefore, they took the birthplace of `Isa as their direction of worship."

Concerning Allah's statement,

"She placed a screen before them;"

This means that she hid herself from them and concealed herself. Then, Allah sent Jibril to her.

"And he appeared before her in the form of a man in all respects."19:17

This means that he came to her in the form of a perfect and complete man. Mujahid, Ad-Dahhak,

Qatadah, Ibn Jurayj, Wahb bin Munabbih and As-Suddi all commented on Allah's statement,

"Then We sent to her Our Ruh,"

"It means Jibril."

"She said: 'Verily, I seek refuge with the Most Gracious from you, if you do fear Allah.'"

This means that when the angel (Jibril) appeared to her in the form of a man, while she was in a place secluded by herself with a partition between her and her people, she was afraid of him and thought that he wanted to rape her. Therefore, she said,

"Verily, I seek refuge with the Most Gracious from you, if you do fear Allah."

She meant, "If you fear Allah," as a means of reminding him of Allah. This is what is legislated in defense against (evil), so that it may be repulsed with ease.

Therefore, the first thing she did was try to make him fear Allah, the Mighty and Sublime. Ibn Jarir reported from `Asim that Abu Wa'il said when mentioning the story of Maryam, "She knew that the pious person would refrain (from committing evil) when she said,

"Verily, I seek refuge with the Most Gracious from

you, if you do fear Allah." He said: "I am only a messenger from your Lord..."'

This means that the angel said to her in response, and in order to remove the fear that she felt within herself, "I am not what you think, but I am the messenger of your Lord." By this he meant, "Allah has sent me to you." It is said that when she mentioned the (Name of the) Most Beneficent (Ar-Rahman), Jibril fell apart and returned to his true form (as an angel). He responded,

`I am only a messenger from your Lord, to provide to you the gift of a righteous son.'

"She said: "How can I have a son..."'

This means that Maryam was amazed at this. She said, "How can I have a son" She said this to mean, "In what way would a son be born to me when I do not have a husband and I do not commit any wicked acts (i.e., fornication)" For this reason she said,

"When no man has touched me, nor am I Baghiyya"

The Baghiyya is a female fornicator. For this reason, a Hadith has been reported prohibiting the money earned from Baghiyy.

"He said: "Thus said your Lord: `That is easy for Me (Allah)...'"

This means that the angel said to her in response to her question, "Verily, Allah has said that a boy will be born from you even though you do not have a husband and you have not committed any lewdness. Verily, He is Most Able to do whatever He wills." Due to this, he (Jibril) conveyed Allah's Words,

"And (We wish) to appoint him as a sign to mankind."

This means a proof and a sign for mankind of the power of their Maker and Creator, who diversified them in their creation. He created their father, Adam, without a male (father) or female (mother). Then, He created Hawwa' (Adam's spouse) from a male (father) without a female (mother). Then, He created the rest of their progeny from male and female, except `Isa. He caused `Isa to be born from a female without a male. Thus, Allah completed the four types of creation (of the human being), which proves the perfection of His power and the magnificence of His authority. There is no god worthy of worship except Him and there is no true Lord other than Him. Concerning Allah's statement,

"And a mercy from Us,"

This means, "We will make this boy a mercy from Allah and a Prophet from among the Prophets. He will call to the worship of Allah and monotheistic belief in Him." This is as Allah, the Exalted, said in another Ayah,

"(Remember) when the angels said: "O Maryam! Verily, Allah gives you the good news of a Word from Him, his name will be Al-Masih, `Isa, the son of Maryam, held in honor in this world and in the Hereafter, and will be one of those who are near to Allah. And he will speak to the people, in the cradle and in manhood, and he will be one of the righteous." 3:45-46

This means that he will call to the worship of his Lord in his cradle and while and adult. Concerning His statement,

"And it is a matter (already) decreed (by Allah)."

This is the completion of Jibril's dialogue with Maryam. He informed her that this matter was preordained by Allah's power and will. Muhammad bin Ishaq said,

"And it is a matter (already) decreed (by Allah)."
This means that Allah determined to do this, so there is no avoiding it.

*"So, she conceived him, and she withdrew with him
to a remote place. And the pains of childbirth drove
her to the trunk of a date palm. She said:*

*"Would that I had died before this, and had
been forgotten and out of sight!"*

JESUS (JUST US)

By George Hopkins

PT.1: FISHING FOR MEN

Why fish for men? Well, it's common knowledge that God talks to man through man so those had to be enlightened to spread the truth. Jesus' teachings were very simple: "Freedom, Justice and Equality; The upliftment of humanity and Love, Peace and Happiness." And he wasn't coined "The Christ" until after he mastered the mystery schools of his time. The definition of Christ is "the anointed one that comes to crush the wicked."

The Bible speaks of a festival that he disappeared in at the age of twelve. Somehow his parents, after a day (or more) of travel, discovered that their son wasn't with them and had to travel back to find him. It is during this time (which is made to seem short), an eighteen-year gap where he attended the mystery school and attained the title "Jesus the Christ." He popped back up on the scene at the age of thirty - which is an important number because most people (if not all) didn't get their calling for prophet hood until their early thirties.

This man's purpose was to introduce a truth to an otherwise uncivilized government rule. The revolutionary of his time. He went against the dictators back then and was executed for it! Even though he healed people (physically, mentally and spiritually); saved people's lives and/or brought them back from death, this wasn't enough to save his own life. An innocent man was tried and convicted on trumped-up charges. The very people who were praising him at the beginning of the week were screaming crucify him at the end of the week, and this was all due to the manipulation of that present regime.

The most important thing to remember about him is he never asked (or ordered) anyone to praise him- which was a testament to his humility. In fact, his exact words were, "if any want to become my followers, let them deny themselves and take up their cross and follow me."

Now the question that protrudes to the forefront is how many of us have denied ourselves? Who is carrying their cross? Are we emphatically following the man we claim to love? I would say, "NO", but also that, the moment we decide to fully walk what he taught we will be tried and hung from a tree just like he was.

"The God of our fathers raised up Jesus, whom ye slew and hanged on a tree."
(King James Bible – Book of Acts 5:30).

PT.2: THE INNER WAR

In light of my above commentary, the statement; *"Peace will never last and war will never end.* "holds true in every facet of the world; every facet; especially within our inner being, as the outer world is a direct reflection of what's going on inside of us. So, it is safe to say that the world is sick because we are sick.

We all go through it even though most do not recognize it. A good sign of the inner war is when we lash out at others as we believe the war/fight must be and can only be fought physically or verbally. There are many more signs but now is not the time. If you can win the internal battle, you'll be well on your way.

Follow my lead...

The late great Jesus; a prophet to some and God to others was undeniably an extraordinary orator. I'm pretty sure we can all agree with that.

Jesus went into the wilderness for forty days and fasted the same amount of time (according to the Bible) and within this journey he encountered the devil!

Three key things transpired here. First; he removed himself from every one by going into the wilderness (though within the wilderness is a jungle). Second; he stayed for forty days and fasted the same number of days. Third and most importantly, he was tempted by the Devil.

The first order of questioning has to be;

"Why did he go into the wilderness in the first place?"

U-N-I-VERSE I: JESUS

"Why fast for forty days?"

"How did the Devil know he would be there?"

Being as though the Bible is 75% allegory and 25% decisive in my opinion, I ask the question:

"What does the wilderness actually represent?"

The wilderness could be a representation of the flesh which would then mean by going into the wilderness, he actually went within *self* (meditation, introspection, reflection, etc.) for forty days. He also abstained from food for forty days. Within this time limit he encountered temptation. As we all know, a dearth of food will make us weak; and any and every time we're in a state of weakness our decision-making becomes distorted. We can easily be swayed. Am I right?

It is no mystery that after fasting - and only after fasting - was Christ Jesus tempted. However, it wasn't the *"devil"* we're taught exist but an internal battle/conflict that he was having and once he mastered/conquered his lower self (the flesh) he became a master!

This is exigent to essay because it teaches us to embrace hardship; to endure the battle with in; to readily and willingly isolate or separate ourselves and meditate at the drop of a dime to find answers. It teaches us that fasting may be hard but that it is also healthy and that even though we may become weak physically due to this physical, mental, and spiritual abstention, if and when we get through the inevitable temptation, we too will become masters of self.

Jesus went into the wilderness by himself and left by himself which can only mean that he was by himself the entire time. This also means that he was battling himself the entire time. Even if you don't believe that what I am saying is true you can't deny the reference of my break down and how it relates to life.

PT.3: F.E.A.R. FALSE EVIDENCE APPEARING REAL

True indeed, it is imperative that we question everything that comes across our table, for fear (false evidence appearing real) of becoming slaves to an erroneous doctrine or form of indoctrination.

"my people are destroyed for lack of knowledge..."
-Hosea 4:6

So, the first thing we must get his knowledge and we get that by studying with an open mind. Especially knowing that the mind is like a parachute - if opened it can save us but if closed it serves us no purpose.

Follow my lead...

According to the book of Mark 8:22, It is said that Jesus cured a blind man at Bethsaida. We take this at face value with no further investigation which is idiotic and shows that it is easy for us to be led in the wrong direction but hard for us to be led in the right direction.

"He took the blind man by the hand and led him out of the village and when he had put saliva on his eyes and laid his hands on him, he asked him, can you see anything in the man looked up and said I can see people but they look like trees, walking. Then Jesus laid his hands on his eyes again and he looked intently and his sight was restored, and he saw everything clearly."
-8:23-25

Now let's look at it...

The man is said to be blind, and Jesus put saliva on his eyes. And laid his hands on him ask him, "can you see anything?" the man replied, "I can see people, but they look like trees, walking."

The first thing that comes to mind, is, was this man really blind? I mean for him to see men as trees, walking, shows and proves that he could in fact see. Am I right? Because if he was truly blind, he wouldn't have seen anything.

Also, what's the deal with the saliva that Jesus put on his eyes? I'm most certain that if anyone had spit in their eyes, they wouldn't be able to see.

The Most Honorable Elijah Muhammad taught that the Bible is 75% allegory and 25% decisive, and it takes a discerning eye/mind to decipher these scriptures.

Looking at things from that standpoint, and using intelligence, I can see that saliva is a liquid that comes out of a person's mouth. The saliva is symbolic to water which is symbolic to wisdom.

Keeping that in mind, let us go back and revisit the scripture. Jesus put saliva in the man's eyes. These eyes are actually the mind's eye and the saliva is wisdom used to *wizen* the dome of the man which gave him sight, or more correctly, helped him see things for what they really were and not what they appeared to be. Remember, to the (spiritually) blind man, men appeared to be trees, up until he received the saliva (wisdom) of the Christ Jesus.

This is our reality today; it isn't just some two-thousand-year-old folk tale. Many of us see but we have no understanding of what we see.

Just as the man saw people as trees, we know and understand that men are as far from trees as the devil is to God. So, what did he really mean? Not that they were tress, but instead they fit the description of such. They were alive (as are trees) but appeared as inanimate wooden figures; contributing nothing to life! Once the man was able to see clearly (as he worded it) he saw men as they were; good, bad, civilized, uncivilized, representations of God or children of the devil, etc.

I hope I aided your spiritual vision the way Jesus did the blind man.

THE LOST YEARS OF SAINT ISSA

The Best of the Sons of Men

Ancient scrolls reveal that Jesus spent seventeen years in India and Tibet.

From age thirteen to age twenty-nine, he was both a student and teacher of Buddhist and Hindu holy men.

The story of his journey from Jerusalem to Benares was recorded by Brahman historians.

Today they still know him and love him as St. Issa. Their 'Buddha'.

In 1894 Nicolas Notovitch published a book called The Unknown Life of Christ. He was a Russian doctor who journeyed extensively throughout Afghanistan, India, and Tibet. Notovitch journeyed through the lovely passes of Bolan, over the Punjab, down into the arid rocky land of Ladak, and into the majestic Vale of Kashmir of the Himalayas. During one of his journeys, he was visiting Leh, the capital of Ladak, near where the Buddhist convent Himis is. He had an accident that resulted in his leg being broken. This gave him the unscheduled opportunity to stay awhile at the Himis convent.

Notovitch learned, while he was there, that there existed ancient records of the life of Jesus Christ. In the course of his visit at the great convent, he located a Tibetan translation of the legend and carefully noted in his carnet de voyage over two hundred verses from the curious document known as "The Life of St. Issa."

He was shown two large yellowed volumes containing the biography of St. Issa. Notovitch enlisted a member of his party to translate the Tibetan volumes while he carefully noted each verse in the back pages of his journal.

When he returned to the western world there was much controversy as to the authenticity of the document. He was accused of creating a hoax and was ridiculed as an imposter. In his defense he encouraged a scientific expedition to prove the original Tibetan documents existed.

One of his skeptics was Swami Abhedananda. Abhedananda journeyed into the arctic region of the Himalayas, determined to find a copy of the Himis manuscript or to expose the fraud. His book of travels, entitled Kashmir O Tibetti, tells of a visit to the Himis gonpa and includes a Bengali translation of two hundred twenty-four verses essentially the same as the Notovitch text. Abhedananda was thereby convinced of the authenticity of the Issa legend.

Translated text:

... He passed his time in several ancient cities of India such as Benares. All loved him because Issa dwelt in peace with Vaishas and Shudras whom he instructed and helped. But the Brahmins and Kshatriyas told him that Brahma forbade those to approach who were created out of his womb and feet. The Vaishas were allowed to listen to the Vedas only on holidays and the Shudras were forbidden not only to be present at the reading of the Vedas, but could not even look at them.

Issa said that man had filled the temples with his abominations. In order to pay homage to metals and stones, man sacrificed his fellows in whom dwells a spark of the Supreme Spirit. Man demeans those who labor by the sweat of their brows, in order to gain the good will of the sluggard who sits at the lavishly set board. But they who deprive their brothers of the common blessing shall be themselves stripped of it.

Vaishas and Shudras were struck with astonishment and asked what they could perform. Issa bade them

"Worship not the idols. Do not consider yourself first. Do not humiliate your neighbor. Help the poor. Sustain the feeble. Do evil to no one. Do not covet that which you do not possess and which is possessed by others."

Many, learning of such words, decided to kill Issa. But Issa, forewarned, departed from this place by night.

Afterward, Issa went into Nepal and into the Himalayan mountains ....

"Well, perform for us a miracle," demanded the servitors of the Temple. Then Issa replied to them*: "Miracles made their appearance from the very day when the world was created. He who cannot behold them is deprived of the greatest gift of life. But woe to you, enemies of men, woe unto you, if you await that He should attest his power by miracle."*

Issa taught that men should not strive to behold the Eternal Spirit with one's own eyes but to feel it with the heart, and to become a pure and worthy soul....

"Not only shall you not make human offerings, but you must not slaughter animals, because all is given for the use of man. Do not steal the goods of others, because that would be usurpation from your near one. Do not cheat, that you may in turn not be cheated ...."

"Beware, ye, who divert men from the true path and who fill the people with superstitions and prejudices, who blind the vision of the seeing ones, and who preach subservience to material things. "...

Then Pilate, ruler of Jerusalem, gave orders to lay hands upon the preacher Issa and to deliver him to the judges, without however, arousing the displeasure of the people.

But Issa taught: *"Do not seek straight paths in darkness, possessed by fear. But gather force and support each other. He who supports his neighbor strengthens himself.*

"I tried to revive the laws of Moses in the hearts of the people. And I say unto you that you do not understand their true meaning because they do not teach revenge but forgiveness. But the meaning of these laws is distorted."

Then the ruler sent to Issa his disguised servants that they should watch his actions and report to him about his words to the people.

"Thou just man," said the disguised servant of the ruler of Jerusalem approaching Issa, *"Teach us, should we fulfill the will of Caesar or await the approaching deliverance?"*

But Issa, recognizing the disguised servants, said, *"I did not foretell unto you that you would be delivered from Caesar; but I said that the soul which was immersed in sin would be delivered from sin."*

At this time, an old woman approached the crowd, but was pushed back. Then Issa said, *"Reverence Woman, mother of the universe,' in her lies the truth of creation. She is the foundation of all that is good and beautiful. She is the source of life and death. Upon her depends the existence of man, because she is the sustenance of his labors. She gives birth to you in travail, she watches over your growth. Bless her. Honor her. Defend her. Love your wives and honor them, because tomorrow they shall be mothers, and later-progenitors of a whole race. Their love ennobles man, soothes the embittered heart and tames the beast. Wife and mother-they are the adornments of the universe."*

"As light divides itself from darkness, so does woman possess the gift to divide in man good intent from the thought of evil. Your best thoughts must belong to woman. Gather from them your moral strength, which you must possess to sustain your near ones. Do not humiliate her, for therein you will humiliate yourselves. And all which you will do to mother, to wife, to widow or to another woman in sorrow-that shall you also do for the Spirit."

So taught Issa; but the ruler Pilate ordered one of his servants to make accusation against him.

Said Issa: *"Not far hence is the time when by the Highest Will the people will become purified and united into one family."*

And then turning to the ruler, he said, *"Why demean thy dignity and teach thy subordinates to live in deceit when even without this thou couldst also have had the means of accusing an innocent one?"*

THE ISLAMIC VIEW OF JESUS
By Abu Amir

In the name of Allâh, The Most Gracious, The Most Merciful. Indeed, all praise is for Allâh, we praise Him, repent to Him, and seek his forgiveness and help. We seek refuge in Allâh from the evil of our own selves and our wicked deeds. Whomsoever Allâh guides, none can lead astray: and whomsoever Allâh leaves astray, none can guide. And I bear witness that none has the right to be worshipped but Allâh alone, and He has no partner; and I bear witness that our Prophet Muhammad is His servant and Messenger. May Allâh, send peace and blessings upon him, his family and his Companions until the Day of Judgement. Ameen.

INTRODUCTION

Whether it be the Talmud of the Jews vilifying Jesus and his mother as an adulteress and bastard child, or the Qur'anic and Ahadith narratives establishing him as a true and just messenger of God or Allah, or the Christian tradition raising him to divine levels, there is indeed a man of flesh and bones being discussed and described and not merely a symbolic figure or mythical folk tale.

Jesus or Isa is mentioned 59 times in the holy Quran. 23 times he is mentioned as Ibn Maryam or "the son of Mary", 11 times he is mentioned as Al-Masih or "the Messiah" and 25 times as Easa (Isa) or Jesus.

From the standpoint of a Muslim (One who submits his will to the Creator of the Universe and all worlds, the source and sustainer and origin of all life forms) Isa (Alayhi Salaam) is of high ranking amongst the Prophets and Messengers of Allah subhanahu wa ta'ala. He is referred to as Ruhullah or the "Spirit of Allah". He is referred to as the "word of Allah" ...

("Be" and it is!)

...as he was sent as a sign to mankind of Allah's omnipotence and sovereignty over creation.

TEACHINGS

Isa alayhi salaam (Jesus, peace be upon him) Describes his earthly mission as a sort of receiving and passing of a prophetic baton or torch in a long line of messengers from Musa alayhi salaam (Moses, peace be upon him) to the paraclete or comforter Muhammad ﷺ. He emphasizes that he has not come to change the law or Torah of Musa (Moses) but to fulfill it and that until the day of Qiyamat (resurrection) not one iota small percentage or jot and tittle (written deen) will be changed. Henceforth' We hear the Prophet Muhammad ﷺ emphasizing that he is simply the seal of the profits, or, the last messenger in a line of messengers from Adam to Nuh (Noah) to Musa (Moses) to Isa (Jesus) May Allah's peace and mercy be upon them all.

The Quran mentions some 25 prophets and or messengers by name and in authentic ahadith there is made mention of over 144,000 messengers and prophets; some with no followers and some with multitudes of followers.

So, what is this message that each messenger of Allah from the prototypical human being and father of humanity, Adam alayhi salaam, to Noah to Job to Seth to Idris (Enoch) to David and Solomon to Isa to the anonymous messengers to the final messenger Muhammad ﷺ taught? Tauhid or:

"The Oneness of God (Allah)"

This message, that there is only one "Source" for all of creation from the smallest atom to the largest mountain and that to be successful in this life and the next, one must submit to the laws of nature stipulated by the creator, has not changed since the beginning of creation. This teaching of oneness leads to balance and harmony and unity. Hence, the poor are to be provided for by the wealthy, each soul should treat and interact with every soul it encounters as it wishes to be treated, that the Lord Your God is One;

The message has not and will not change. The law, the Torah, the Way...it does not change and every messenger sent to rekindle the consciousness of humans who have been duped into slumber by ignorance and evil reminds mankind of this law.

Isa alayhi salaam was inspired by Allah like every messenger who seeks seclusion and is awakened internally to remind mankind of the balance, of the laws that lead to order. He was sent to remind man of the mirage-like nature of the material world's ephemeral existence and inform him of the reality of his soul; the infinite nature of his soul's existence and the end for those who do good as well as those who do bad. The message never changed, only the physical applications of the law which adapt to time, culture, eon, etc. which still, in essence remain the same and as firmly intact as the cycles of the celestial bodies orbiting the sun in our solar system.

A man blinded by his own ignorance can however perceive an enlightened being as a divine entity for whom worship and praise are due. This is merely admiration exceeding the limits. Once Isa alayhi salaam realized that the devotion of some of his followers had evolved into worship of his physical being he admonished them, rebuking all forms of praise and worship and reminding them of his human limitations.

*...yet you refuse to come to Me to have life. **I do not accept glory from men**, but I know you, that you do not have the love of God within you...."*

-Berean Study Bible (John 5:40-42)

"And Jesus answered and said unto him, get thee behind me, Satan: for it is written, thou shalt **worship the Lord thy God, and him only shalt thou serve.** *"*
-Luke 4:8

"Jesus therefore said to them again, (Salamun Alaykum)"Peace be with you; as (Elah/Allah) **the Father has sent me, I also send you.** *"*
-John 20:21

"And I fell at his feet to worship him. And he said unto me, 'See thou do it not: I am thy fellow servant and of thy brethren that have the testimony of Jesus: **Worship Elah (God): for the testimony of Jesus is the spirit of prophecy.** *"*
-Revelation 19:10

According to Ibn Ishaq Allah's Messenger ﷺ was seated in the mosque with Walid ibn Mughirah. Nadr bin Harith also came and sat down and some of the Quraysh were already there. Allah's Messenger ﷺ began to speak and Nadr interrupted him. Allah's Messenger ﷺ silenced him with his reply and recited the verse of al-Anbiya

"Surely you and what you worship besides Allah Shall be fuel for Hell. You shall go down it." -21:98.

Then as he departed, Abdullah ibn Sahmi came and sat down. Walid ibn Mughirah told him that the Prophet ﷺ had recited the verse (21:98) but Nadr could not say anything. Abdullah said, *"I would have silenced him. If all those we worship besides Allah are fuel for fire then we worship the Angels, the Jews worship Uzayr, the Christians worship Easa."*

Walid and the others applauded him, imagining that he had presented an irrefutable argument. When the Prophet ﷺ was told of this, he said, *"Whoever wishes that he should be worship besides Allah, will go to hell along with his worshippers. These people worship the devil and the devil commands them to worship the jinn."*

The verse was revealed:

"Surely, those for whom the goodly reward has preceded from Us, they shall be kept the far away from it - they shall not hear a whisper of it, and they shall abide in that their souls desired."
-21:102

Thus *Sayyidina Easa, *Sayyidina Uzayr and the ulama and aesthetics whom the people worship will be 'kept far away from Hell'.

*Sayyidina is an Arabic Term meaning 'Our Master'. It is a term used by Allah subhanahu wa ta *ala in al-Qur'an Karim to describe his al-anbiya (the Prophets).*

" And remember when Allah will say on the day of resurrection; O Isa, son of Maryam, did you say unto men worship me and my mother as two Gods besides Allah? He will say, glory be to you! It was not for me to say what I had no right to say! Had I said such a thing, you would surely have known it. You know what is in my inner self though I do not know what is in yours, truly, you, only you, are the all knower of all that is hidden and unseen."
Surah Al-Maeda (5:116)

BURIAL NEXT TO MUHAMMAD ☐

Tirmizi reported that it is mentioned in the Torah that Sayyidina Easa will be buried with Sayyidina Muhammad ﷺ. Abu Mawdud stated that there is space in the shrine for one grave.

Ibn Kathir has stated that "Easa's demise will take place in Medina, where his Janazah (Burial) Salaat will be performed. He will then be buried beside the Holy Prophet Muhammad ﷺ."

BIBLE VERSION OF ISA'S TEACHINGS

In Jeremiah chapter 8 verse 8 we read:

"How do you say, we are wise, and the law or Torah of the Lord is with us? Low, certainly in vain made he it or (revised it), the pen of the scholars is in vain."

Which coincides with The Quran's assessment of the "Revised" Scripture:

"Then woe to those who write the Book with their own hands and they say, 'This is from God (Allah).' To purchase with it a little price! Woe to them for what their hands have written and woe to them for that they earn thereby."
-2:79

"And verily, among them is a party who distort the Book with their tongue (as they read), so that you may think it is from the Book, but it is not from the Book, and they say, 'This is from God, but it is not from God; and they speak a lie against God while they know it."
-3:78

So, in the current version of the Bible we see the Bible questioning its own "current" authenticity. In verse nine of Jeremiah, after condemning the scribes for changing and or forging the words of the book and attributing them to Allah (God) we then see the shifting of the Lord's Covenant from the Israelites to "others" who will inherit their wives, and land (fields), because even the "priests" have been corrupted. This too is corroborated by the Glorious Qur'an in which Allah states what translates as:

"So, because of their breach of their covenant, We cursed them, and made their hearts grow hard. They change the words from their (right) places and have abandoned a good part of the message that was sent to them. And you will not cease to discover deceit in them, except a few of them."
-5:13

Hence, when the Holy Quran speaks of the Injeel (Gospel of Jesus not the teachings of Paul which often contradict the teachings of Jesus) and Torah (books of Moses) it is not speaking of the Bible Commissioned by Constantine the Great at the Council of Nicaea in 325AD nor the King James revision completed in 1611.

The Quran is speaking of the actual oral and written versions taught by Isa (Jesus) and Musa (Moses) to their disciples and followers directly from the mouths and actions of these two great Prophets (May Allah's peace and mercy and blessings be upon them both). Portions of Isa's original Injeel was discovered at Nag Hammadi, Egypt in 1945 and a year later near the Dead Sea which borders Israel, the West Bank and Jordan. Unlike a majority of the Pauline version of the New Testament, these two important pieces of the "Apocrypha" were written by disciples and brothers of Isa (Jesus) such as Barnabas,

Thomas and others who had actually "seen", "lived" and "worshipped" with the man we refer to as Jesus.

NAG HAMMADI

In 1945, a group of peasants were digging for fertilizer near the village of Nag Hammadi in Egypt when they serendipitously uncovered a jar containing 13 leather-bound manuscripts buried sometime in the late fourth century. When the manuscripts came to the attention of scholars of history their significance was recognized immediately. They contained fifty-two tractates, mostly "heretical" writings of Gnostic Christians (Christian ascetics enthralled by spiritual mysticism).

Although originally composed in Greek, the writings were in Coptic, or, ancient Ethiopian translation. Many of them had been previously known by title only. Today these writings are known as the Nag Hammadi library.

THE COPTIC APOCALYPSE OF PETER

The Coptic Apocalypse of Peter is a very interesting treasure from the Nag Hammadi discovery because it tells a story of a crucifixion strikingly similar to the crucifixion narrative found in the Quran.

In Surah An-Nisa' [4] ayah (verses) 156-158 we read:

"Is that they rejected faith; that they ordered against Mary a grave false charge; that they said in boast, 'We killed Christ Jesus the son of Mary, the Apostle of Allah' but they killed him not, nor crucified him, but so it was made to appear to them, and those who differ therein are full of doubts with no certain knowledge, but only conjecture to follow, for Surely, they killed him not but Allah raised him up unto himself and Allah is exalted in power, wise."

Peter gives a vision of Jesus never dying on a cross but an appearance of Jesus being killed. We read:

Jesus said to Peter 'And they will put themselves to shame. But me they cannot touch. And you, O Peter, will stand in their midst. Do not be afraid because of your cowardice. Their minds will be closed. For the invisible one has opposed them."

*When he had said those things, I saw him **apparently** being seized by them. And I said, **'What am I seeing, O Lord? Is it you yourself who they take? And are you holding on to me? Who is this one above the cross, who is glad and laughing? And is it another person whose feet and hands they are hammering?'***

The Savior said to me? "He whom you see above the cross, glad and laughing, is the living Jesus. But he into who's hands and feet they are driving the nose is his physical part, **which is the substitute. They are putting to shame that which is in his likeness.** *But look at him and look at me."*

But I, when I had looked, said, 'Lord, no one is looking at you. Let us flee this place.'

But he said to me, "I have told you, leave the blind alone. And noticed how the y do not know what they are saying. For the son of their glory, instead of my servant, they have put to shame."

The Acts of John (Nag Hammadi) describe Isa as a *Docetic* or a spirit that never possessed a real flesh and body.

In the letter of Peter, Peter is urging James to pass along the sermons that he has enclosed and his letters to those who are worthy to receive them. The overtone of Peter's letter is a concern that his teachings not be corrupted by those who have a *'different and distorted understanding of the truth'.* Peter describes this distorter of truth as 'the man who is my enemy' and this man is known to be none other than the Apostle Paul, who taught that salvation came to all people, Jew and Gentile regardless of whether or not they followed the Law of Moses.

However, his views stood in stark contrast to the views of Jewish Christians like the Ebonite.

Peter opens his letter to James with 'Peace be with you' or in the Syriac Aramaic that they spoke *'As Salaamu alaikum'* and says,

'The same way as Moses handed over his office of a teacher to the seventy. Wherefore also the fruit of his caution is to be seen up to this day. For those who belong to his people preserve everywhere the same rule in their belief in the one God and in their line of conduct, the Scriptures with their many senses being unable to incline them to assume another attitude.

'Rather they attempt, on the basis of the rule that has been handed down to them, to Harmonize the contradictions of the Scriptures, if haply someone who does not know the traditions is perplexed by the ambiguous utterances of the prophets.

On this account they permit no one to teach unless he first learns how the Scriptures should be used. Wherefore there obtain amongst them one God, one law, and one hope....

'...and indeed, some have attempted, while I am still alive, to distort my words by interpretations of many sorts, as if I taught the dissolution of the law and, although I was of this opinion, did not express it openly. But that may God forbid!

For to do such a thing means to act contrary to the law of God which was made known by Moses and was confirmed by our Lord in its everlasting continuance. For he said: **the heaven and earth will pass away, but one jot or one tittle shall not pass away from the law.'**

This he said that everything might come to pass. But those persons who, I know not how, allege *that they are at home in my thoughts wish to expound the words which they have heard of me better than I myself who spoke them. To those whom they instruct they say that this is my opinion, to which indeed I never gave a thought.*

But if they falsely assert such a thing while I am still alive, how much more after my death will those who come later venture to do so?"

ISA'S PASSING OF THE TORCH

Ibn Kathir writes: "Allah said about the last prophet of the Banu Isra'il (Isa) that he said: *'Surely, I am the messenger of Allah to you confirming that which was revealed before me in the Torah, and giving glad tidings of a messenger who shall come after me whose name shall be Ahmad.'"* (61:6)

The Gospel of Barnabas mentions Muhammad by name while the Bible we have today gives references to and descriptions of him. The Injeel (Gospel of Isa) gives tidings of the coming of Farqalit [Greek 'Paraclete' see: John 14:16] which refers to Muhammad. The term paraclete is often misrepresented as the "Holy Spirit" but the Holy Spirit had been present before Jesus, during his life and after his ascension.

Muqatil Ibn Hayyan said that Allah, the Exalted, the Glorious, revealed to Isa:

"Preach my Commands, Oh son of the chaste woman. I have created you without a father and thus made you a sign for the worlds only that you worship Me. Preach to the people of Suran in Syrian language and tell them that I am true, Ever Existing. Confirm the unlettered Prophet who rides the camel, wears the armor, the crown and the sandals, the possessor of the staff. He has curly hair, broad forehead, joined eyebrows, large eyes, long eyelashes, black eyes, curved nose, distinct cheeks and a dense beard. Drops of perspiration of his face like pearls. A pleasant scent of musk emanates from him. He has a fine neck, and fine, soft hair grows from his chest to navel, no other hair on his stomach. He overshadows others with him and walks as if descending from a height. He has no male child."

CONCLUSION

His birth was a reminder of the Creator's will, might and omnipotence. His mission was the same as the messengers that came before him; to deliver the glad tidings of Paradise for those whose good deeds outweigh their bad ones on the Day of judgement and a warning of eternal punishment for evil-doers who fail to repent and amend their ways. He was a deliverer of the Torah or "The Law"; the law that governs all things in existence and a reminder that man must align himself with this law to be successful. This law is submission to the will of the Creator.

His mission was to remind that *"The Lord Your God is One."* or la ilaha ilallah. His mission was to reestablish the societal balance that had been disrupted by the rich who saw fellow human beings not as people to love and fellowship with but to exploit and take advantage of. Thus, giving alms or paying zakat and prayer or salaat were at the forefront of his teachings.

He was very clear that all men had the power to ascend to his spiritual plane and therefore forbade the worshipping of anything other than the source of creation. Allah azza wa jal. He despised materialism and he shunned evil. He advocated fasting and seclusion.

We should not reduce him to mere myth and it appears he did not want us to elevate him to divinity but rather we should deal with the man who was a humble messenger and whose brother Thomas referred to as *"a servant of God"* and seek to emulate every positive character trait he projected. May Allah accept my humble offering. May he protect me from the evil in my own soul and may he protect me from the punishment of the grave. Ameen

"Christ the son of Mary was no more than a messenger; many were the messengers that passed away before him. His mother was a woman of truth. They both had to eat their (daily) food. See how Allah doth make His signs clear to them; yet see in what ways they are deluded away from the truth!"
(5:575)

The Honorable Elijah Muhammad and Master Fard Muhammad

Mirza Ghulam Ahmad (as) wrote in his book, Izala Auham (Removal of Doubts):
"I certainly believe in the rising of the sun from the West; that said, it has been disclosed to me in a vision that the meaning of the rising of the sun from the West is that the Western countries, which, from ancient times, have been enveloped in the darkness of disbelief and error, will be illumined by the sun of truth and will partake of Islam"

(Below) Masjid Maryam Chicago, Illinois Est. 1972

The Roza Bal, Rouza Bal, or Rozabal is a shrine located in the Khanyar quarter in downtown area of Srinagar in Kashmir, India. The word roza means tomb, the word bal mean place. Locals believe a sage is buried here, Yuz Asaf,[2] alongside another Muslim holy man, Mir Sayyid Naseeruddin.

The shrine was relatively unknown until the founder of the Ahmadiyya movement, Mirza Ghulam Ahmad, claimed in 1899 that it is actually the tomb of Jesus.[2][3] This view is maintained by Ahmadis today, though it is rejected by the local Sunni caretakers of the shrine, one of whom said "the theory that Jesus is buried anywhere on the face of the earth is blasphemous to Islam."

Some Muslims believe that Jesus will be buried alongside Muhammad at the fourth reserved tomb of the Green Dome in Medina.

Image of Jacob's Vision of a Ladder on which angels ascend and
descend in heaven.

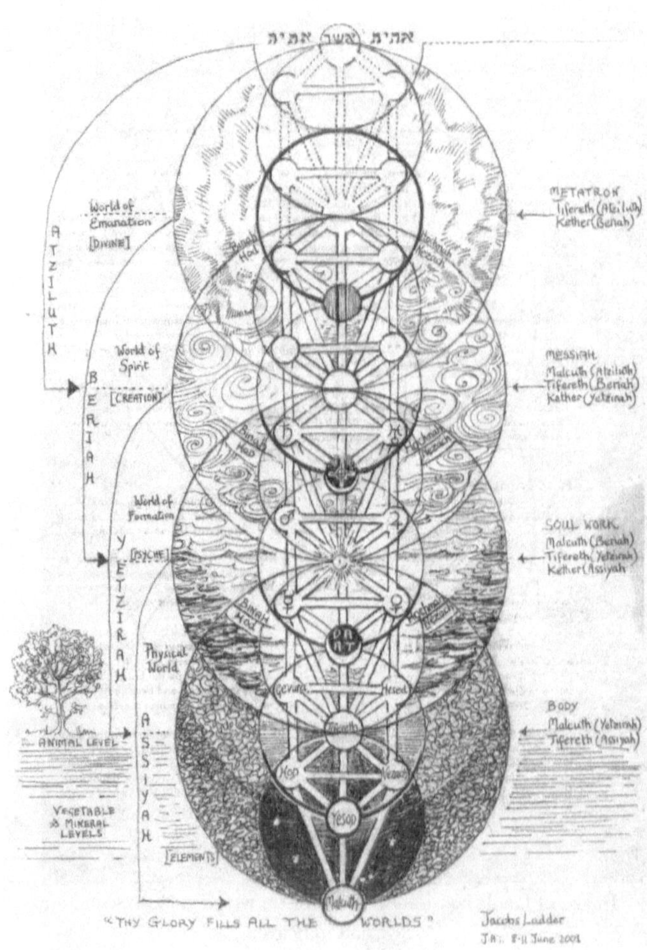
אהיה אשר אהיה
World of Emanation
[DIVINE]
ATZILUTH
World of Spirit
[CREATION]
BERIAH
World of Formation
[PSYCHE]
YETZIRAH
Physical World
ASSIYAH
ANIMAL LEVEL
VEGETABLE & MINERAL LEVELS
[ELEMENTS]
METATRON
Tifereth (Atziluth)
Kether (Beriah)
MESSIAH
Malcuth (Atziluth)
Tifereth (Beriah)
Kether (Yetzirah)
SOUL WORK
Malcuth (Beriah)
Tifereth (Yetzirah)
Kether (Assiyah)
BODY
Malcuth (Yetzirah)
Tifereth (Assiyah)
Hod
Netzach
Tifereth
Yesod
Malkuth
"THY GLORY FILLS ALL THE WORLDS"
Jacobs Ladder
JA. 8-11 June 2001

The Building Blocks of Life

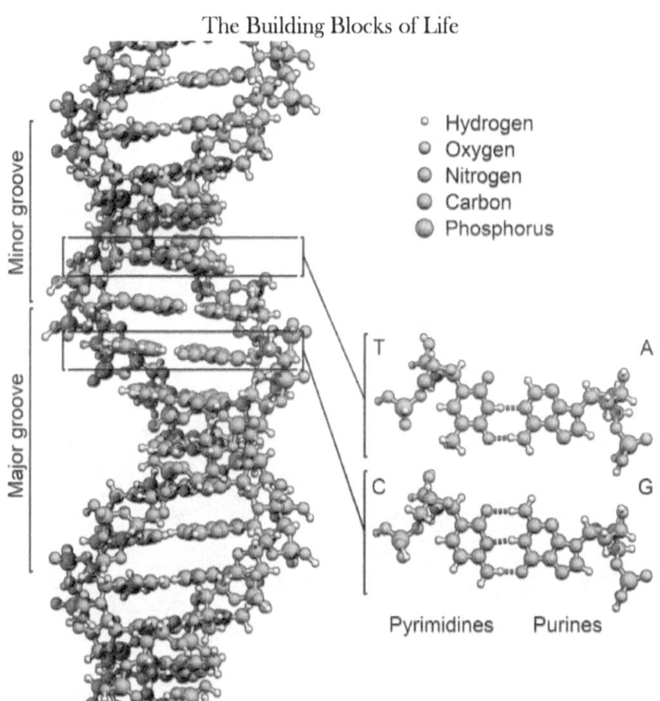

The Kingdom of God within
"The Holy of Holies"

John 2:15–16. And Jesus went into the temple of God, and cast out all them that sold and bought in the temple, and overthrew the tables of the money changers, and the seats of them that sold doves, and said unto them, It is written, My house shall be called the house of prayer; but ye have made it a den of thieves.

1.This little child Jesus when he was five years old was playing at the ford of a brook: and he gathered together the waters that flowed there into pools, and made them straightway clean, and commanded them by his word alone. 2 And having made soft clay, he fashioned thereof twelve sparrows. And it was the Sabbath when he did these things (or made them). And there were also many other little children playing with him. 3 And a certain Jew when he saw what Jesus did, playing upon the Sabbath day, departed straightway and told his father Joseph: Lo, thy child is at the brook, and he hath taken clay and fashioned twelve little birds, and hath polluted the Sabbath day. 4 And Joseph came to the place and saw: and cried out to him, saying: Wherefore doest thou these things on the Sabbath, which it is not lawful to do? But Jesus clapped his hands together and cried out to the sparrows and said to them: Go! and the sparrows took their flight and went away chirping. 5 And when the Jews saw it they were amazed, and departed and told their chief men that which they had seen Jesus do.

-The Infancy Gospel of Thomas

Then will Allah say: "O Jesus the son of Mary! Recount My favour to thee and to thy mother. Behold! I strengthened thee with the holy spirit, so that thou didst speak to the people in childhood and in maturity. Behold! I taught thee the Book and Wisdom, the Law and the Gospel and behold! thou makest out of clay, as it were, the figure of a bird, by My leave, and thou breathest into it and it becometh a bird by My leave, and thou healest those born blind, and the lepers, by My leave. And behold! thou bringest forth the dead by My leave. And behold! I did restrain the Children of Israel from (violence to) thee when thou didst show them the clear Signs, and the unbelievers among them said: 'This is nothing but evident magic.'

إِذْ قَالَ ٱللَّهُ يَـٰعِيسَى ٱبْنَ مَرْيَمَ ٱذْكُرْ نِعْمَتِي عَلَيْكَ وَعَلَىٰ وَٰلِدَتِكَ إِذْ أَيَّدتُّكَ بِرُوحِ ٱلْقُدُسِ تُكَلِّمُ ٱلنَّاسَ فِي ٱلْمَهْدِ وَكَهْلاً وَإِذْ عَلَّمْتُكَ ٱلْكِتَٰبَ وَٱلْحِكْمَةَ وَٱلتَّوْرَىٰةَ وَٱلْإِنجِيلَ وَإِذْ تَخْلُقُ مِنَ ٱلطِّينِ كَهَيْـَٔةِ ٱلطَّيْرِ بِإِذْنِي فَتَنفُخُ فِيهَا فَتَكُونُ طَيْرًا بِإِذْنِي وَتُبْرِئُ ٱلْأَكْمَهَ وَٱلْأَبْرَصَ بِإِذْنِي وَإِذْ تُخْرِجُ ٱلْمَوْتَىٰ بِإِذْنِي وَإِذْ كَفَفْتُ بَنِي إِسْرَٰٓءِيلَ عَنكَ إِذْ جِئْتَهُم بِٱلْبَيِّنَٰتِ فَقَالَ ٱلَّذِينَ كَفَرُواْ مِنْهُمْ إِنْ هَٰذَآ إِلَّا سِحْرٌ مُّبِينٌ

"The rose gives the bees honey..."

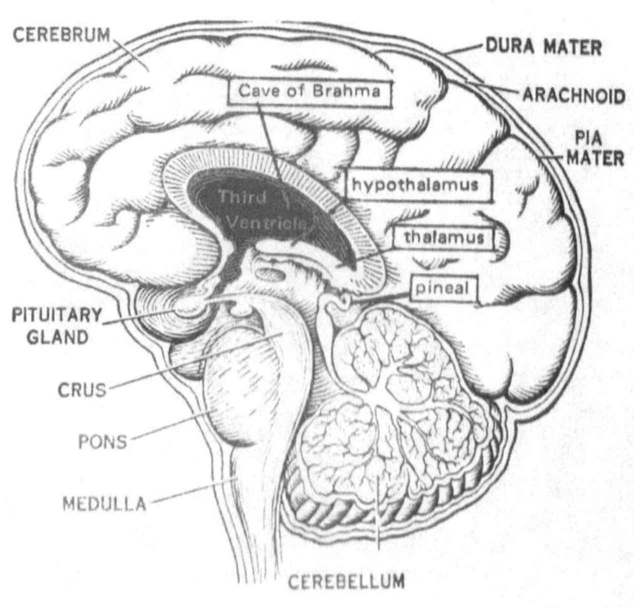

CEREBRUM
DURA MATER
Cave of Brahma
ARACHNOID
PIA
MATER
hypothalamus
Third
Ventricle
thalamus
pineal
PITUITARY
GLAND
CRUS
PONS
MEDULLA
CEREBELLUM

U-N-I-VERSE I: JESUS

"And [mention] when Jesus, the son of Mary, said, "O children of Israel, indeed I am the messenger of Allah to you confirming what came before me of the Torah and bringing good tidings of a messenger to come after me, whose name is Ahmad." But when he came to them with clear evidences, they said, "This is obvious magic." -Quran 61:6

Louis Tiffany, Holy Spirit or Dove Window (1895-1900), First Presbyterian Church of Springfield, Illinois.

A few Muslim commentators, such as David Benjamin Keldani (1928), have argued the theory that the original Koine Greek used was periklytos, meaning 'famed, illustrious, or praiseworthy', rendered in Arabic as Aḥmad (another name of Muhammad), and that this was substituted by Christians with parakletos.

The Nicene Creed

The purpose of a creed is to provide a doctrinal statement of correct belief among Christians. The creeds of Christianity have been drawn up at times of conflict about doctrine: acceptance or rejection of a creed served to distinguish believers and heretics. For that reason, a creed was called in Greek a σύμβολον, symbol on, which originally meant half of a broken object which, when fitted to the other half, verified the bearer's identity. The Greek word passed through Latin symbolum into English "symbol", which only later took on the meaning of an outward sign of something.

The Nicene Creed was adopted to resolve the Arian controversy, whose leader, Arius, a clergyman of Alexandria, "objected to Alexander's (the bishop of the time) apparent carelessness in blurring the distinction of nature between the Father and the Son by his emphasis on eternal generation".[13] Emperor Constantine called the Council at Nicaea to resolve the dispute in the church which resulted from the widespread adoption of Arius' teachings, which threatened to destabilize the entire empire. Following the formulation of the Nicene Creed, Arius' teachings were henceforth marked as heresy.[14]

The Nicene Creed of 325 explicitly affirms the Father as the "one God" and as the "Almighty," and Jesus Christ as "the Son of God", as "begotten of [...] the essence of the Father," and therefore as "consubstantial with the Father," meaning, "of the same substance" as the Father; "very God of very God." The Creed of 325 does mention the Holy Spirit but not as "God" or as "consubstantial with the Father." The 381 revision of the creed at Constantinople (i.e., the Niceno-Constantinopolitan Creed), which is often simply referred to as the "Nicene Creed," speaks of the Holy Spirit as worshipped and glorified with the Father and the Son.

Lucius Caecilius Firmianus Lactantius was an early Christian author who became an advisor to the first Christian Roman emperor, Constantine I, guiding the emperor's religious policy as it developed during his reign. His work De Mortibus Persecutorum preserves the story of Constantine's vision of the Chi Rho before his conversion to Christianity. The full text is found in only one manuscript, which bears the title, Lucii Caecilii liber ad Donatum Confessorem de Mortibus Persecutorum.

The bishop Eusebius of Caesaria, a historian, states that Constantine was marching with his army (Eusebius does not specify the actual location of the event, but it is clearly not in the camp at Rome), when he looked up to the sun and saw a cross of light above it, and with it the Greek words "(ἐν) τούτῳ νίκα" ("In this, conquer"), a phrase often rendered into Latin as in hoc signo vinces ("in this sign, you will conquer").

At first, Constantine did not know the meaning of the apparition, but on the following night, he had a dream in which

Christ explained to him that he should use the sign of the cross against his enemies. Eusebius then continues to describe the Labarum, the military standard used by Constantine in his later wars against Licinius, showing the Chi-Rho sign. The accounts by Lactantius and Eusebius, though not entirely consistent, have been connected to the Battle of the Milvian Bridge (312 AD), having merged into a popular notion of Constantine seeing the Chi-Rho sign on the evening before the battle.

The phrase appears prominently placed as a motto on a ribbon unfurled with a passion cross to its left, beneath a window over the Scala Regia, adjacent to the equestrian statue of Emperor Constantine, in the Vatican. Emperors and other monarchs, having paid respects to the Pope, descended the Scala Regia, and would observe the light shining down through the window, with the motto, reminiscent of Constantine's vision, and be reminded to follow the Cross.

The Kingdom of Portugal had used this motto since 1139, according with the legend in Lusíadas.

"Ἐν τούτῳ νίκα"

Constantine and The Council of Nicaea

"In hoc signo vinces"
"In this sign thou shalt conquer"

The original Nicene Creed was first adopted at the First Council of Nicaea in 325. In 381, it was amended at the First Council of Constantinople. The amended form is also referred to as the Nicene Creed, or the Niceno-Constantinopolitan Creed for disambiguation.

The Nicene Creed is the defining statement of belief of Nicene or mainstream Christianity and in those Christian denominations that adhere to it. The Nicene Creed is part of the profession of faith required of those undertaking important functions within the Orthodox and Catholic Churches.

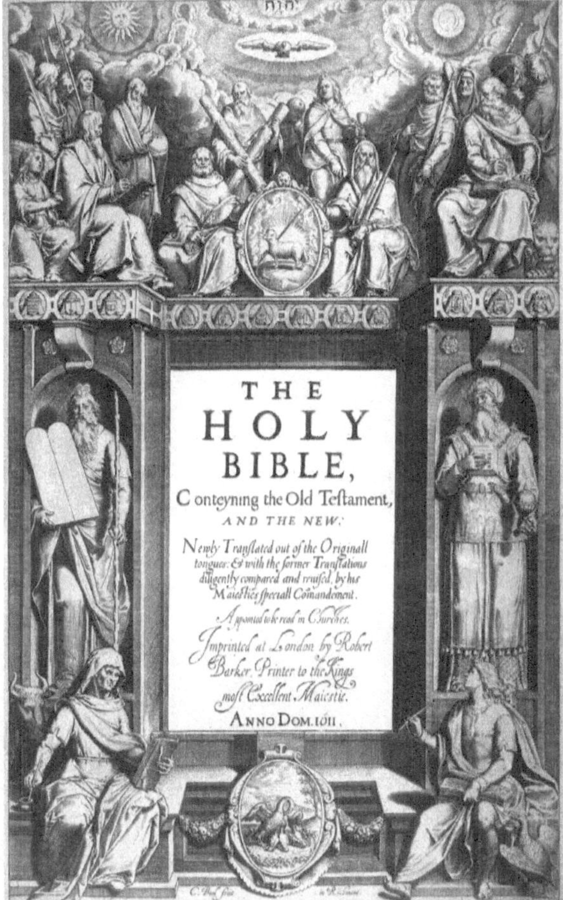

King James Bible 1611

    The King James Version (KJV), which was published in 1611 at the behest of King James I of England. James was also the head of the Church of England, and therefore approved the new English translation of the Bible, which was also dedicated to him.

    As far as its authors; Richard Bancroft, the archbishop of Canterbury—was notable for having the role of overseer of the project but the actual translating (writing) of the KJV was done by a committee of forty-seven scholars and clergymen over the course of many years.

King James 1 of England

THE TRUE HISTORY OF JESUS
By Saleem

There is plenty to unpack, unravel, illuminate and elucidate when exploring the personage of the one we today refer to as Jesus. There is the hunt for the *historical* Jesus which remains a mystery only to those unfamiliar with what those sworn to secrecy refer to as *the Mysteries.* Those acquainted with this deliberately hidden information are well aware of the empirical evidence of the true history of Jesus, his travels, his whereabouts, death and remains.

There is the *mythologized* Jesus, utilized for the establishment of a new order amongst Church, State and Paganism, and the juxtaposing of his story with the countless crucified saviors before him, most notably his resemblance to Mithra.

To complete the triad you have the mystical, *esoteric Jesus,* the ubiquitous spirit capable of being channeled by any illumined mind. By the will of Allah, I strive to touch upon all three aspects for it is the consolidation of these three facets that unfold and reveal *The True History of Jesus.*

As with his death, Jesus's birth is shrouded in mystery, the knowledge of which was only known to a select few, his immediate family being amongst those *in the know.* According to the New Catholic

Encyclopedia and the Encyclopedia of Early Christianity *"The exact date of Christ's birth is not known."* For now, we will establish it was not in the month of December.

Shepherds were *"living out of doors and keeping watches in the night over their flocks."* (Luke 2:8) *The book Daily Life in the Time of Jesus* notes that flocks lived in the open air from *"the week before the Passover [late March]"* through mid-November. It then adds: *"They passed the winter under cover; and from this alone it may be seen that the traditional date for Christmas, in the winter, is unlikely to be right, since the Gospel says that the shepherds were in the fields."*

This is merely one myth the true Jesus of two-thousand years ago has suffered from and his true story has never been revealed to the public. Not in the Gospel, The Apocrypha or the Nag Hamadi scrolls, nowhere will you find the unadulterated story of Jesus and his ministry outside of a few Essene and Anti-Nicene sects, some of which do still exist to this day.

Not until the 1930s did the History of Jesus become open to the public because it wasn't until this time that the prophecy of his return began to manifest and fulfill itself. When a mysterious traveler from the East appeared and began teaching the son of a Southern Preacher the true History of not only Jesus, but the Original Asiatic man in general.

There are facts and artifacts still concealed in the secret vaults in the hidden possession of certain enlightened fraters. Those few *Knights Templars* who had been initiated into the arcana of the Essenes, the Nazarenes, the Johannites and the Druses in the Holy Land knew and those who had followed in their line of secrecy know.

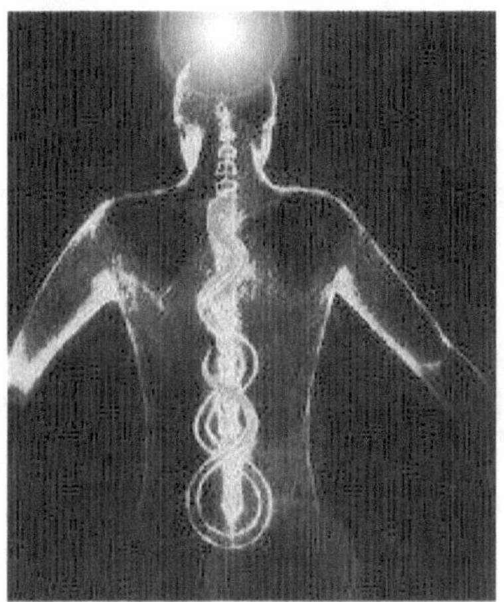

The builders of the church (institution) often revealed their intellectual inadequacies and lack of empiricism when formulating a doctrine so full of contradictions, hyperbole and discrepancies. They had no intimate relationship with Jesus or his followers and evidently based their writings on mythical folklore and exaggerated heresy.

According to the most common notion, Jesus was crucified at the age of thirty-three. This, in fact, is only as true as a Mystic's conviction in sacred geometry and arithmetic. In regards to his physical age however, we find more controversy.

"They (Gnostics), however, that they may establish their false opinion regarding that which is written, 'to proclaim the acceptable year of the Lord,'

maintain that he preached for one year only, and then suffered in the twelfth month. They are forgetful of their own disadvantage, destroying his whole work, and robbing him of that age which is both necessary and more honorouble than any other; that more advanced age, I mean, during which also as a teacher He excelled all others. For how could He have had His disciples, if He did not teach? And how could He have taught, unless He had reached the age of a Master? For when He came to be baptized, He had not yet completed his thirtieth year, but was beginning to be about thirty years of age (for thus Luke, who has mentioned His years, has expressed it: 'Now Jesus was, as it were, beginning to be thirty years old,' when He came to receive baptism); and, (according to these men,) He preached only one year reckoning from His baptism. On completing His thirtieth year He suffered, being in fact still a young man, and had by no means attained advanced age.

Now, that the first stage of early life embraces thirty years, and that extends onward to the fortieth year, every one will admit; but from the fortieth and fiftieth year a man begins to decline towards old age, **which Our Lord possessed while He still fulfilled the office of a Teacher, even as the Gospel and all the elders testify**; those who were conversant in Asia with John, the disciple of the Lord, (affirming) that John conveyed to them that information. And he remained among them up to the time of Trajan. Some of them, moreover, saw not only John, but the other apostles also, and heard the very same account from them, and

bear testimony as to the (validity of) the statement.
Whom then should we rather believe? Whether such
men as these, or Ptolemæus, who never saw the
apostles, and who never even in his dreams attained
to the slightest trace of an apostle?"
-St. Irenæus, Bishop of Lyons (AD 180)

Godfrey Higgins, writes, in regards to this passage by Irenaeus: "...the Doctrine of the crucificion was a *vexata questio* among Christians even during the second century. *"The evidence of Irenæus,"* he says, *"cannot be touched. On every principle of sound criticism, and of the doctrine of probabilities, it is unimpeachable."*

The omitting and deleting of such statements always has been and always will be intentional.

"If the disciples themselves related that Jesus lived to advanced age in the body, why has the mysterious number 33 been arbitrarily chosen to symbolize the duration of his life? Were the incidents in the life of Jesus purposely altered so that His actions would fit more closely into the pattern established by the numerous Savior-Gods who preceded Him?"
-Manly P. Hall

In an attempt to rectify any issues with accurately "chronicling" the life of Jesus we realize there are simply too many ulterior motives involved, none of which belong to the man himself.

"It is by no means improbable that Jesus Himself originally propounded as allegories the cosmic activities which were later confused with His own life. That the **Χρίστος**, *Christos, represents the solar power reverenced by every nation of antiquity cannot be controverted. If Jesus revealed the nature and purpose of this solar power under the name and personality of Christos, thereby giving to this abstract power the attributes of a god-man, He but followed a precedent set by all previous World-Teachers. This god-man, thus endowed with all the qualities of Deity, signifies the latent divinity in every man. Mortal man achieves deification only through at-one-ment with this divine self. Union with the immortal Self is therefore "saved". This Christos, or divine man in man, is man's real hope of salvation – the living Mediator between abstract Deity and mortal humankind."*

-Manly P. Hall

"They teach the immortality of the soul and esteem that the rewards of righteousness are to be earnestly striven for. Yet, it is

their course of life better than that of the other men and they entirely addict themselves to husbandry."
-Jewish Historian, Josephus, on the Essenes

There is an entire history, credulous and incredible, that supplants the gospels as recorded by the authors known as Matthew, Mark, Luke and John. Records of early Christian visitation can be found in the hills of Tibet and the vaults of Ceylon monasteries. Also, there are tales of his sojourning in Greece as well as India (Kashmir in particular). If and when he visited these places is debatable, however, if there is even an inkling of truth attached to these rumors of Jesus' travails and travels, and there usually is, then there may also be truth to the possibility of his exposure to Asiatic and pagan Greek wisdom.

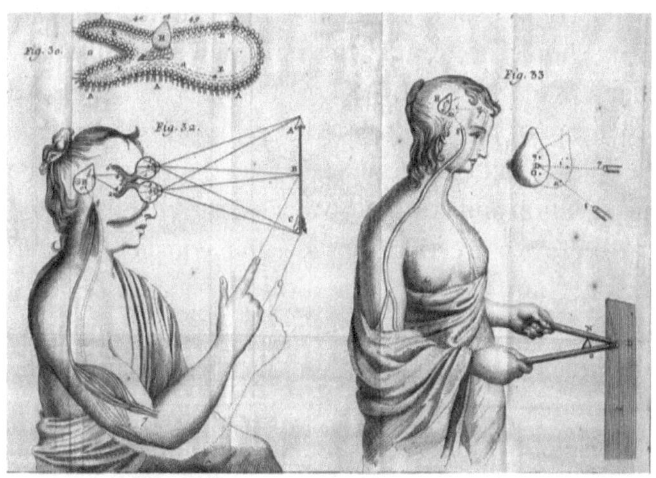

As the portrait of Jesus comes into focus, we see a normal man with a divine mission. Why a divine mission? Because he set his heart upon doing the will of Allah (God) since a baby in the cradle and never wavered even until old age. To fulfill the mission, he set his heart upon, the mission instilled in the subconscious of every human, he takes what he is taught in Kemet (Cairo, Egypt) and travels a northeastern direction to give and receive light; North to Give, East to receive. As with all prophets and seers in or around Jews or Jerusalem he must flee persecution. In the Bible his biggest enemies seem to be Pharisees and Sadducees who sought to silence or kill any one who strayed from the letter of the law and questioned arbitrary order. He must also return to be murdered by them.

Traveling eastward, Greece, India, he is further enlightened, like the christened Buddhas before him. Espousing the most ancient wisdom from the schools of Kemet, the enlightening, universally aligning teachings of the East and the homogenized doctrine of the Jews, modified by Moses for the people of Yakub (Jacob), Jesus then returns to face the persecution and crucifixion that every free thinker and advocate for freedom, justice and equality suffered before him. *Calvary*, the deciding factor between those willing to remain enslaved to the opinionated doctrines of others or free themselves from the quagmires of dogmatism and proclaim their "oneness" with the Creator (seen as the Father and Ancient of Days to the Ancients). Those willing to declare with conviction;

"There is no way to the father but through me," properly translated, *there is no way to the supreme architect but through enlightenment." For truly, there is no way to achieve oneness with the multi-universe (sentient life) and Universal Mind (Allah) than through CHRISTening. (Heavenly Anointment)"*

The few disciples he could influence he taught and showed well, instructing them (awakening and quickening in them) the power to heal the blind and curse the wicked. The ability to regulate the "sacred secretions" of the pineal gland through meditation and "faith", thus germinating the seeds of divinity within all beings. [Matt. 10:1, Mark 6:7, Luke 9:1, Luke 10:19, Rev. 2:6]. In leaving he never tells his disciples to worship him as the Christ but to be Christ' themselves. His fate remains a mystery unless one has studied the teachings of Master Fard Muhammad as taught by the Honorable Elijah Muhammad and or is able to piece the puzzle created by man's flawed history accounting system. One of our biggest flaws is sensationalizing and over-romanticizing history. As Manly Hall stated, *"at one time an asset, the "miracles" of Christianity have become its greatest liability. Supernatural phenomena, in a credulous age interpolated to impress the ignorant, in this century have only achieved the alienation of the intelligent."*

Maulan Muhammad Ali in his Tafsir of Surah 3:44 says, "...The whole history of Mary and Jesus was enveloped in darkness until the Holy Quran announced their right position as two righteous servants of God and rejected both extreme views, the Jewish view that Jesus was conceived in sin and was illegitimate and the Christian view that he was God or the Son of God who had entered Mary's womb. He was only what the prophet described him to be in his controversy with the *Najran* deputation when he said to them:

*"'Do you not know that Jesus was **conceived** by a woman in the manner in which all women **conceive**? Ten she was **delivered** of him as women are **delivered** of their children? Then he was fed as children are fed. Then he ate food and drank water and answered the call of nature (as all mortals do)?" The deputation replied to all these questions in the affirmative, on which the Prophet said: "Then how can your claim (that he was God or the Son of God) be true?'*

"The Prophet's clear argument which even the Christian deputation could not question settles the matter that Jesus was conceived in the ordinary manner and that Mary became a wife and mother in the ordinary way."

So how then do we finally reconcile the historical, metaphorical, and *"miraculized"* story of Jesus, Isa, Yashua, etc.? Through the teachings of the Honorable Elijah Muhammad, a man that offered $10,000 to any man who could prove just one of his statements false.

"I have taken up Jesus and his teaching because after Master Fard Muhammad taught me, I saw quickly that this man's (Jesus') history was the world's most misinterpreted and misunderstood."

According to the teachings of the Honorable Elijah Muhammad, Joseph and Mary had known each other since they were very little children and had attended school together.

"They always loved each other. They agreed to marry when very small children going to school. When Joseph and Mary grew to the age of fifteen and seventeen, Joseph went to Mary's father and told him that he loved Mary and Mary loved him and that they wanted to get married. Joseph was a poor man, a commoner. He also was a carpenter by trade; not a first - class carpenter, but one of those building little rough houses in South Europe.

Mary's father war rich and had plenty of property. Also, her grandfather was rich. So, Mary's father objected to her marriage to Joseph, a poor carpenter. Mary loved Joseph and told her father that if she did not marry him, she would not marry anyone else. When the old man heard this, it made him sick, because no father wants his daughter to be an old maid. Joseph gave up thinking about marring Mary and went and married another girl. Mary's father would go out and bring in very rich young men for his daughter in hopes of his daughter taking a liking to some of these young men, but she did not.

Joseph owned a little house, a small piece of land, and was also rearing up a family. He soon was the father of six children by this girl that he married and was married for six years. Mary's father was a big wealthy man and a great architect.

The old man had to go away to another town, about twenty – five miles from his home and set up a temple, but before he left home, he called Mary and side to her, " Now daughter, I have to go away and I will be gone for about three days. Now you must take care of everything until I return. You must feed and water all of the livestock. " (The old man had plenty of cattle.) He also said to Mary: "I want you to wear my old clothes and put on some false whiskers so that the people when they see you going down to feed the stock, they will think you are me an no one will say anything to you.

After the man leaves on his journey, a great dust storm comes up, which made it very dark. This kind of storm comes up twice a year. During this storm, Mary be – comes frightened and sends a message to Joseph, telling him that her father had gone away and that he would be away for three days. That she was left to feed and water all of the livestock and that she was alone and was a little afraid. She asked Joseph to come over and help her to take care of everything and the he would be company for her. Joseph tells his wife that he has to go off on a little business. He came over to Mary's house and she tells him of the costumes of her father what she would wear when she goes out to feed the livestock. Joseph said, "Give me the old man's costume and when anybody sees us, they will think I am the old man with you." Joseph gets the team ready and Mary gets in with him to feed the livestock. The next day Joseph comes back and stays with Mary and helps her to feed and take care of the livestock.

On the third day, Mary's father came back and asked his daughter, " How 's everything?" She said, "All right, Father." So, after three months the old man noticed Mary was growing larger than usual. He said to Mary, "What is the matter with you fattening so fast? Are you eating more than you should?" Mary said, "No, Father, I am not eating more than I should." The old man called Mary again and said "Mary, there is something the matter with you. Look how large you are. What are you doing?" Mary no longer could hide it, so she told him the truth. She said to him, "Father, do you remember when you went off to set up a temple and were gone for three days some time ago and left me to attend the livestock? Well, I was afraid because of a dust storm and sent for Joseph to come over and help me. Well, he came and did help me feed the stock." The old man said, "Yes, and he fed them, too." She said to him, "Do you remember Joseph and me were trying to get you to let us marry when we were young and you would not allow us?

Now this is what happened and I shall send for Joseph and let him tell you for himself." An old woman had already been to Joseph and told him that the child Mary was carrying was his child and the child *would be a wonderful child, a prophet and teacher. She told him that the child was his, but Joseph denied it. But this woman knew because she was a prophetess, or what we call a medium. She told him his name was in the Holy Qur'an and that he should take care of him and that the authorities would seek the child's life.*

Joseph went and told Mary the good news that the old woman had told him concerning the child. Then Joseph went to Mary's father. By this time the old man was really sick because Mary was growing larger and larger every day. He was in bed and had just pulled all of the whiskers out of his chin. Joseph said to the old man: "I will tell you about the whole thing, then you can kill me right here. Mary and I always loved each other, even when we were going to school and promised to marry when we were old enough. I came to you and asked you for Mary's hand but you would not give her to me. Now this is what happen. I am the father of the child that Mary is carrying but remember that the child is going to be a great man, a prophet and teacher. His name can be found in the history of the Holy Qur'an. Now remember, you must not tell anyone about this because of the authorities. If they knew this, they would kill the child. So, keep your mouth shut about this. Now you can kill me if you want to. I have told you the truth."

The last month of Mary's pregnancy drew near but the authorities
did not know what time the child would be born. So, the same medium that told Joseph of the child also

went and told the authorities but no one knew which day the child would be born, although they knew the week. The authorities had a committee to go around and take a census of all the pregnant women and when they would give birth. This committee checked everyone who was pregnant. When the week came that Jesus would be born, the authorities had given orders that all boy babies born in that week should be killed. Joseph came and took charge of Mary himself. All men over in Asia are taught how to take care of their wives and perform this duty of their wives. In case they don't care for a wet nurse, the girls are also taught about midwifery when still very young.

Mary's father had plenty of livestock and ordered stalls to be built. Joseph took one of the stalls and filled it in all around the inside and put a bed in the center for Mary (this is the manger that the Bible speaks about where Jesus was born). To look at the stalls from the outside, it appeared as if it was just a stall filled with hay for the animals. In this way he concealed Mary from the authorities. He brought her food and water every day. In that week when the authorities were expecting Jesus to be born, they sent the committee around to those who had a list of who would give birth that same week. When they came to Mary's father's house, they knocked on the door and the old man went and met them. They said to him," Where is the woman who gave birth this week?" Doesn't she stay here?" The old man said "What woman?" They said, "The one that was with child." He said, "Oh, I know what you are talking about. The woman that was here some time. She was only visiting here," the old man said. She has returned to her home in Egypt. " They said, "Oh, she was just visiting?" The old man said, "Yes that is all. "Joseph came daily with water and food to Mary and to see the baby. He fooled his wife by telling her that he had a job building a house. This was how Jesus was born.

After a few days and as soon as Mary was able, Joseph told her that he would have to send her and the child away to his people down in Egypt. Joseph gets himself and Mary all prepared to leave for Egypt. At a certain time of the year in Palestine where they were living a dark cloud comes up. When this kind of cloud comes up, one cannot see so well. It was the kind of cloud that Joseph waited for. Under the darkness of this stormy cloud, Joseph mounted Mary and her baby on the back of an ass. Away they went for the camel station. The ass (or donkey) knew the route under the darkness of the storm cloud. (the camel station is where camels are for hire.) Over at home (Asia) we have camel stations. They are called mail camels. The camel is very sensitive and a fast runner. The mail camels go from forty to fifty miles per hour and it was one of the mail camels they hired. From Jerusalem to Cairo, Egypt was about six hundred miles. The camel knew the route well. He travels it daily. Nothing is necessary except to tell the camel where you want him to go and he will go right direct to the place. They have comfortable seats on their backs. Joseph had provided Mary with money *and everything. So he mounted her and Jesus on the mail camel, and told the camel to take them to Cairo, Egypt, as fast as possible. Off they went and Joseph and the ass returned home.*

Mary took the child Jesus into Egypt where his people lived at Cairo. When Jesus was about the age of five, his mother put him in school. A very intelligent child in school he was. He finished school at the age of twelve. He took courses in astronomy, geometry, and all other branches of sciences.

There was an old man who had learned that Jesus was born and he wanted to see Jesus and must see him at any price. He knew what Jesus was going to be and how the authorities were going to seek him and try to destroy him. This old man wanted to get to Jesus and teach him how to stay out of the way of those who would seek to kill him. This old man's lesson that he wanted to teach Jesus was about the Radio in the Head." This old man knew it. It is called Radio in the Head because with this great wisdom you can hear as distinctively as one can hear over a wire or like the present radio. The man actually hears what other men think or say in their heads. Jesus must know this wisdom.

The old man goes and gets on the road that Jesus uses to come and go to school, but he must see him alone. The first day the old man watched the boys come from school with Jesus. When Jesus got out of school, he and the other boys would be together. The old man could not get to talk with Jesus. He goes back and says, "I will come back tomorrow. I will get right in front of him and fall across his path." The next day the old man looking for Jesus saw him and the other boys coming alone. Jesus was talking about his lessons and was teaching another boy his algebra.

The old man pretended to the boys that he was looking for a certain street number and Jesus heard him asking the boys the number, and the number he called was just a few doors from his home. Jesus said, "I know where you want to go." The other boys went home and left the old man and Jesus together. The old man said to Jesus, "I heard you talking about algebra. You are a very smart boy. I have a grandson and he is studying the same course. I would like for you to come over and teach him his lesson as you could help him a lot." Jesus said, "Yes, I will." This just suited Jesus. He liked to teach someone something.

The old man did not have anyone for Jesus, nor did he care about the number that Jesus was taking him to. He said to Jesus, "Do you know who you are and what your name is?" Jesus said, "Yes my name is Jesus, and I don't know any more than that. I believe that I will be a great man someday." "Yes," said the old man. "Your name is in the Qur'an. Have you seen it there? You are that man. I am not looking for a number. I was looking for you. I have something that I want to teach you because as soon as you finish school, you will leave home and you are going to Europe to the Adamic Civilization to teach. Also, you must know how to take care of yourself because they will try to kill you."

The next day the old man goes and gets himself onto the path Jesus uses. It just so happened that Jesus was all alone this day, walking behind his classmates, so the old man was very glad. He walked with Jesus toward his mother's house and this time he gives his first lesson about receiving messages (Radio in the Head). It was easy for Jesus to learn anything. He learned it within three days. Then the old man said to him, "Now you can go anywhere you want to. You can take care of yourself; nobody can harm you now unless you let them. "

After Jesus finished school, he said to his mother that he was going to Europe. His mother did not want him to go. She gave him everything that she thought he wanted. At the age of fourteen his mother could not keep him any longer. She gave him two thousand dollars the day he left to go to Europe.

Jesus had not gone but a few blocks before he had given the two thousand dollars away that his mother had given him. Then he left Cairo and walked six hundred miles to Europe from Jerusalem in his bare feet, teaching as he traveled. If he got hungry, he would not eat your bread unless you let him work for it. He would wash his own clothes in a lake or someplace where he could find water. He never ate a meal without working for it but once, and that was one day when he walked until his feet was sore and swollen. He stopped at this woman's house that is in your Bible, by the name of Martha.

Luke 10:38 says: "Now it came to pass, as they went, that he entered into a certain village: and a certain woman named Martha received him into her house." She bathed his feet and prepared him some food. This is not the only time that he was at her home.

Jesus had a hard time trying to teach the people Freedom, Justice, Equality. He taught anywhere and everywhere that he could get the people to listen to him. He never had over twenty -five or thirty listeners at one time. He would start teaching at a place and before he could finish the authorities would be on their way after him. But he always stayed tuned in on them. As soon as they would start after him, and would get near him, he would dismiss his congregation and run for his life.

They never would have killed Jesus if he had given himself up because he always knew what they were going to do before they could start. The basis of Jesus' teaching were Freedom, Justice, and Equality, or to sum it up, " The Sermon on the Mount." Jesus taught "As a man thinketh in his heart so is he." Jesus' teachings were designed to show the Adamic Race that they would eventually self - destruct because of their own evil doings and not from some avenging God in the sky! Repent of your evil and be saved. The Beatitudes teach the principles and rules of the Kingdom of God.

Jesus 'constant cry was, " I come not to destroy the law but to fulfill it." To fulfill means to be in actuality. The authorities chased Jesus until he became discouraging to them. They named him Christ (which means troublemaker) because every time they would get there, he would be gone. Jesus taught twenty - two years on the run. Until, at last he read and understood that he was to soon: two thousand years to soon! Jesus was two thousand years ahead of time for the teaching of freedom, justice, and equality.

It was raining that day. He got himself a standing place under an awning of a Jew's store and began teaching those who were stopping under the awning out of the rain. The Jew who owned the store got angry with Jesus about teaching in front of his store and asked him to leave. Jesus told him to let him teach under his awning, that the people he was teaching would buy something from him. But the Jew insisted that the people were keeping those that might buy something from seeing his goods in the showcase. But Jesus said, " I don't care. If I leave here, I will lose my people who are listening to me. " He had about twenty - five there. Jesus never had over thirty - five listening to him at any one time. The multitudes that the Bible speaks of only heard about Jesus from word of mouth. The Jew said, " You are the man called Jesus that the authorities are looking for a long time." Jesus said, "Yes I am the one and you can call them anytime you want to." The Jew goes in and sends for the authorities, informing them that Jesus was teaching right in front of his store. "Come right over, right away, and you will catch him," he said.

Two officers (policeman) were sent to the scene. Now there was a reward out for Jesus. The reward was twenty-five hundred in gold (not thirty *pieces of silver) if he was brought in dead, and fifteen hundred dollars if he were brought in alive.*

So, these two officers raced down to the old Jew's store where said Jesus was. Each of the officers wanted the reward. Jesus knew that they were coming and made no effort to leave because he intended for them to kill him. The two officers rushed up to him, one on each side of him, each man trying to be the first to lay hands on Jesus so he could get the reward because the reward would only be paid to one man.

Both of the men (officers) came near, touching Jesus at the same time. The one on the right said Jesus was his man. The one on the left said, "No, he's mine. I was the first to touch him". The other said, "I was the first. Jesus seeing them arguing and disputing about him, said to them. "Will you let me help you settle this argument and tell you who was the first man to put his hands on me?" The two officers agreed to let Jesus settle it for them. "The man on the right side touched me one tenth of a second before the one on the left." So, the one on the left lost the reward. The one on the right who won took Jesus and started on toward the jail. They got down the street and the officer said to Jesus," Listen I will only get fifteen hundred dollars in gold for carrying you in alive, but if I carry you to them, they are going to kill you anyway, so why not let me kill you and make the twenty -five hundred dollars as I am a poor man and have a wife and family to care for?" Jesus knew that he would be killed but did not care.

So, Jesus agreed with the officer that he would let him kill him so he could get the bigger reward. These officers were armed with a knife. The kind of knife that the hand would not slip either way on it like the present hunting knives. Jesus and the officer walked down the street where a Jew's store was vacant with boards nailed over the front to secure the safety of the store. The officer said, "Right here is all right. Now put your back against the wall and your hands up."

Jesus was not afraid. He was a brave man. He put his hands up and straightened himself and put his hands straight across the storefront flat against the boards.

The Honorable Elijah Muhammad taught much more on the history of Jesus and the most intriguing unknown fact was the $6,000 dollar fee, elected *"Christians"* could pay to see the still embalmed body of the Jesus of 2,000 years ago.

As sure as the sands subside and continue to reveal more of the sphynx and its mysteries, the mystery of Jesus will continue to come into the light until it is in full view and until we realize that as the brother George Hopkins elocutes: "Jesus is Just-Us".

He was a man, with an honorable doctrine and mission. His personage was used to create the Commercial Industry known as "The Church", but his teachings weren't lost on those with ears to hear, eyes to see and hearts to comprehend. Internalized, he left a message that enlightened and immortalized a many of men, humble pious carpenters, fisher, farmers and shepherds.

And still, his story doesn't end there. For we must now look into the Biblical and Quranic as well as Hadith and Apocryphal writings that promise His return.

- Surah 43:61
- Revelation 7:1-3

Both are debated however and for many good reasons. For example, the *hu ´* in the 61st verse of the 41st surah is taken to mean *Jesus* but also *The Quran*. The verse in Revelation has a much deeper meaning than a man from 2,000 years ago being resurrected, this is in reference to a man with the same spirit as the Isa from 2,000 years ago who came to usher in the Kingdom of God. What is the Kingdom of God? Sublime peace and enlightenment and oneness with the world soul and Universal mind. It's Islam (Peaceful submission). How does it rule the earth? When the multitude of society is in tune with their inner savior who guides them to eternal gardens of peace with eternal rivers (wisdom) flowing through them, and they begin to share symbiotically these waters and this divine state of conscious awareness and peace. That is the Kingdom of God in Heaven (The Purified mind) descending down to the Earth.

So has a man or men come, who fit the description of the one promised to rise in the last days? Let's take a look at Ahadith.

Dhee Mukhammar related this Hadeeth from the Prophet ﷺ; *"You will make a truce of peace with Rome; you and they will conquer an enemy from behind them. you will be safe and you will achieve spoils. Then you will descend in a fertile soil that has many mounds in it. A man from the Romans will stand, raise the cross, and say, 'victory is for the cross.' A man from the Muslims will rise and kill him. at that point, the Romans will betray their agreement and there will be massacres. They will gather for you, advancing toward you with 80 banners, and with each banner there will be 10,000."*

Anyone familiar with History will see the Story of Saladin and Richard the Lionheart of Jerusalem, although we see Constantine also proclaimed "victory is for the cross".

Elijah Muhamad's mission was to make things clear and to provide proof for those mind's that take more than blind faith to be persuaded. An example of this can be found in a lesson addressing this moment in history when Christianity transformed from a body of knowledge a few ascetics followed to the official religion of the state. Elijah was asked *"Why did we take Jerusalem from the Devil? How long ago?"* His answer, Because *one of our righteous brothers, who was a prophet by the name of Jesus, was buried there. He uses his name to shield his dirty religion, which is called Christianity; also, to deceive the people so they will believe in him. Jesus' Teaching was not Christianity. It was Freedom, Justice and Equality. Jerusalem is in Palestine (Asia Minor). Jerusalem is a name given by the Jews, which means founded in peace, and it was first built by the original man, who was called Jebus; also, Salem and Ariel. We took the city from the Devils about seven hundred fifty years ago."* To quote Micheal Muhammad Knight, *"The Lessons are presented as Fard Muhammad's examination of his student, Elijah Muhammad; while some sections are older than others, copies of Lost Found Muslim Lesson No. 2 date the exam as February 20, 1934."*

If we take the date of the lessons into consideration we see just how accurate his teachings were and enlightening they were to the lost sheep, the lost tribe of Shabazz also referred to as the lost tribe of Israel, many of whom had lost all memory of their history, culture, religion and language.

"Saladin spent over two decades of his life fighting the crusaders and the year 1187 CE would bring him the greatest triumph of his career."
-Syed Muhammad Khan

Jerusalem: called also Salem, Ariel, Jebus, the "city of God," the "holy city;" by the modern Arabs el-Khuds, meaning "the holy;" once "the city of Judah"

(2 Chronicles 25:28). This name is in the original in the dual form, and means "possession of peace," or "foundation of peace." The dual form probably refers to the two mountains on which it was built, viz., Zion and Moriah; or, as some suppose, to the two parts of the city, the "upper" and the "lower city." Jerusalem is a "mountain city enthroned on a mountain fastness" (comp. Psalm 68:15,16; 87:1; 125:2; 76:1,2; 122:3). It stands on the edge of one of the highest table-lands in Palestine, and is surrounded on the south-eastern, the southern, and the western sides by deep and precipitous ravines."

-Easton's Bible Dictionary

"Circle 7" Noble Drew Ali, Moorish Science Temple

Salahuddin (Saladin and the Christians of Jerusalem)
Alphonse-Marie-Adolphe de Neuville - François Guizot. Public
Domain

This new Christ would clear away all clouds of
confusion, curse the synagogue of Satan, tell many

who claim to follow Jesus "I never knew you" and according to hadith;

The messenger of Allah ﷺ said, *"Eesa ibn Maryam will be a fair judge and a just Imam in my nation; he will crush the cross and kill the swine. He will remove the jizya and leave the Sadaqah (charity; this is because wealth will be abundant at that time; and so goats and camels will not be sought after by a charity-tax collector). Also, mutual hatred and rancor will be lifted (from the people). Poison will be removed from all that have poison in them, so that a baby will place his hand in the mouth of a snake, but it will not harm him. a child will drive away a lion, but it will not harm him and a would will be among sheep as if it is their dog."*

Elijah's core message was the destruction of the false religion built around the name of Jesus whose sign was *the cross*. He adamantly admonished the eating of *swine*. Through his dietary instructions on *"How to eat to Live"* he was able to *rid his followers of the poison* in the hearts, minds and bodies. Indeed, Elijah Muhammad was a *fair judge* (Both Malcolm X and Muhammad Ali were suspended, your status didn't matter) and a just Imam. He often opened his lectures with "I am Elijah Muhammad, the teacher of Freedom, *Justice* and Equality..."

More hadith, "He (The Dajjal or Anti-Christ) will stay on the earth for a period that Allah wills, and then Eesa ibn Maryam will *come from the West*, confirming Muhammad upon his *millah* (his creed and Shariah (Laws of Islam)), and he will kill the dajjal.

Then all that remains is the arrival of the hour." (Ahmad). Those familiar with the teachings of the Honorable Elijah Muhammad will recognize his seal immediately in these writings.

"Coming from the west, he indeed *confirmed Muhammad* (peace be upon him) in his *new creed and laws* (study of the Sunnah is Imperative as well as upholding the 5 pillars of Islam).

It was reported that Eesa (Isa) would return and *"will remain on the earth for 40 years – a just Imam and a fair ruler."*

Elijah's "Christening" began with Fard's departure in 1934-1935 and ended in 1974-75 with his death, approximately 40 years he ministered with the spirit of Christ that had descended from the clouds (confusion) with two angels (messengers) and beads of sweat (water is wisdom) dripping from him.

The story of Jesus would never have been a mystery to the one who was to return as the Jesus of the modern era. Elijah asked, *"Can the Devil fool a Muslim?"* Fard replied, *"No, not nowadays."* Abu Hurairah asked, *"O messenger of Allah, will his (devil) doubt-provoking, specious arguments harm me?"*Muhammad replied, *"No, you are a person who is a Muslim, and he is a man who is a disbeliever."*

"Iblees (Satan) will fall down in prostration, calling out, "My God, order me to prostrate to whomsoever you wish." The devils will gather to him and say, "O our leader, to whom are you turning for help? He will say, "I only asked my Lord to give me respite until the Day of Resurrection, and the sun has

indeed risen from its West. This is the well-known time." The devils will become visible on earth..."

Indeed, the *Devils* and their works became visible to the world and the *Man of Sin* [2 Thess. 2:3] was revealed when Jesus, that shining sun, rose in the West. For more truth on Jesus, his mission and his ministry please look towards the teachings of the Honorable Elijah Muhammad and the Honorable Minister Louis Farrakhan for further detailed reading.

"...Incrustations...have attached themselves to the body of Christianity during the centuries. The popular mind itself has been the self-appointed guardian and perpetuator of these legends, bitterly opposing every effort to divest the faith of these questionable accumulations. While popular tradition often contains certain basic elements of truth, these elements are usually distorted out of all proportion. Thus, while the generalities of the story may be fundamentally true, the details are hopelessly erroneous. Of truth as of beauty it may be said that it is most adorned when unadorned. Through the mist of fantastic accounts which obscure the true foundation of the Christian faith is faintly visible to the discerning few a great and noble doctrine communicated to the world by a great and noble soul."

-Manly P. Hall

THE END

AND THE BEGINNING